Print ISBN: 978-1-7770816-0-7
ePub ISBN: 978-1-7770816-1-4

Research and Development in New Brunswick

A History of the Research and Productivity Council

LAURIER L. SCHRAMM

And

ERIC COOK

"An important part of the task of [RPC]… is to exercise careful judgement in choosing fields of research which have a good probability of increasing productivity in existing or new industry… [and that] the government expects assistance to New Brunswick industry in the application of available technology, and the development of new techniques of direct application in the province. Such service will, without doubt, contribute significantly to increased production and productivity and a better standard of living for us all. Expansion, change, new developments — these are becoming the order of the day in New Brunswick."

Louis J. Robichaud
Premier of New Brunswick, 1965 [1]

"The aim of RPC is to help companies use what they've got to answer problems they face. It is easy for a company to solve a production problem 'radically' but few companies, even the larger ones, have the time or are willing to take the risk. Most companies the size we have in New Brunswick can't afford a major turnover. We have got to find for them a way of doing their job quickly and efficiently in the context of the machinery they already have."

Dr. Claude Bursill in 1979 [2]
Executive Director, RPC

"Innovation has to be second nature to us all, whether we work here at RPC or anywhere in the province of New Brunswick."

Bernard Lord
Premier of New Brunswick, 2002 [3]

"the result validates the deeds"

CONTENTS

FOREWORD

When I was asked to write the forward to the History for the Research and Productivity Council my initial reaction was "sure"… that will be easy. I have worked in academia for over 30 years, teaching marketing and acting in a variety of administrative roles ranging from director of a research centre myself to dean of a faculty. Research is an activity I am well acquainted with and writing…. well that is what I do on a daily basis.

During my first month at RPC I visited each department, spoke with each department head, studiously studied my Board materials and plagued Eric with a million questions on the operation, the strategy, the mandate and the role of the Board, and more specifically, the Chair. Suffice it to say that my first round of visits left me overwhelmed with information, impressed by the professionalism and dedication of the RPC team and in awe of the amazing work going on behind the unassuming walls of a building I had been curious about since my UNB student days.

Having served nearly 4 years now as the Chair of the Board of RPC, and sharing bi-weekly calls with the CEO for updates and strategic discussions, I was confident that I had an idea what would be in this book and what I would write to interest you, the reader, in what Eric and his colleague Laurier had to say about RPC. Instead I was riveted by what I was reading. In fact, I have read this book now a total of three times, each time learning something new about the organization I am so intimately involved with. And now, as I sit in front of my computer to tell you why this book is worth a read, I am not sure where to start – the list of reasons is long – very long indeed.

RPC is probably one of the best kept secrets of our province. The non-descript, unassuming building on the corner bordering the UNB Fredericton campus, sandwiched between residence and government garages gives no indication of the intricate laboratories, the high-tech equipment or the brain trust at work behind those walls. A Crown Corporation that is self-sustaining, providing value to a diverse set of stakeholders with a diverse set of needs. Whether it is ensuring water safety for the general population, perfecting a robotic arm for a New Brunswick manufacturer with clients world-wide, or securing highest level of accreditations for scientific testing,

RPC strives to be best in class.

This book is appropriately structured to showcase the evolution of RPC from "the scientific arm of government" to a true partner in using science and technology to support innovation and growth in New Brunswick businesses, support government requirements and share its innovation and research capabilities around the world. The more I read, the more impressed I became with agility of the organization to adapt to the changing, and often chaotic, environment in which it was operating. From reorganizing and establishing subsidiary companies to ramping up contract work to account for changes in, and the eventual elimination of, government contributions to operating costs.

My attention was captivated by the sheer expanse of projects RPC has been, is and has the potential to be involved in. Work like mould and water testing were what I was expecting to read about. Even machine learning was not really a stretch although the concept itself causes me great trepidation which I am reminded of the odd time my watch answers a question that hasn't even been asked yet. But who would have guessed that "Skunked," that spray bottle of lifesaving antidote that is available from your vet for those late-night encounters your dog may have with Pepe Le Pew, is on those shelves because of a curious mind at RPC?

This book reads like a set of Russian Tea Dolls. Every time you think you have come to the end of finding something new, there it is – yet another surprising fact about a world class lab, with state-of-the-art equipment, and bright, creative and inquisitive minds. Things you have read about or wondered about – "Tunagate" or the "Bricklin" or how authorities know that the moose meat in a freezer came from multiple animals - gain a whole new perspective and come to life in the pages of this book.

Engaging and informative, this book is written with respect and devotion by two individuals brought together by a common passion for discovery through science and an obvious appreciation for the role of RTOs in the economic and social prosperity of our region. Their painstaking research and intimate knowledge of one of New Brunswick's best kept secrets builds the story of RPC supporting the importance of recording history much like the Coat of Arms designed for RPC showing black and white symbolizing ink on paper.

Dr. Shelley Rinehart, Chairperson
Research and Productivity Council

PREFACE

In 2020, the New Brunswick Research and Productivity Council (RPC) reached its 58th year of continuous operation. This book is mostly organized into eras, as there have been four eras of distinctly different leadership and strategy.

In what we have termed *The Building Years*, 1962-1969, RPC came into being, with people, technologies, tools, and infrastructure. By 1969, RPC was ready to deploy its services to industry.

In *The Growing Years*, 1970-1983, these capabilities were put to work, and expanded. By demonstrating an ability to quickly deliver practical results for industry problems, results that helped New Brunswick companies to thrive and grow, RPC built-up a critical mass of loyal clients and growing revenues fuelled rapid expansions in staff and capabilities. By 1983, RPC had matured and had taken its place as one of Canada's leading Research and Technology Organizations (RTOs).

In *The Commercial Years*, 1983-2004, RPC became more business oriented and tested how far it could stretch into market-pull commercial operations. Although revenue growth slowed during this era, RPC made great strides in expanding the services it could offer to companies. These ranged from assisting entrepreneurs and small- and medium-sized enterprises (SMEs) with virtually all aspects of product development, commercialization, and production, to assisting large- and multi-national companies with scale-up engineering, pilot- and demonstration-plant verification of technologies and operational assistance. RPC also demonstrated its world-class skills by engaging with companies around the world.

In *The Entrepreneurial Years*, 2004-2020, RPC responded to the trends toward globalization, shortened product life cycles, and rapidly changing economic and political environments. In doing so, the organization became increasingly agile by anticipating and responding to change, offering diverse business services, and emphasizing collaborative efforts where possible. These efforts were so successful that RPC was driven beyond risk mitigation and into another phase of strong growth.

Perhaps the most notable feature of RPC's history, and one that spans all of its organizational eras, is the translation of its employees' skills and talents into technological innovation on the part of RPC's clients, which has propelled New Brunswick companies, small and large, to decades of commercial successes that have helped strengthen New Brunswick's economy and quality of life.

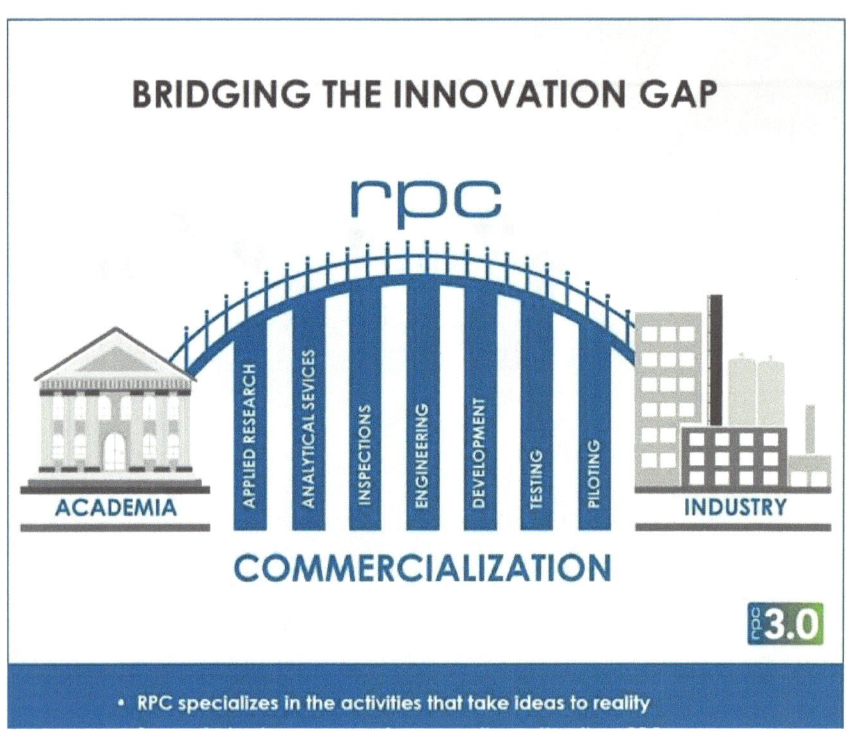

ACKNOWLEDGMENTS

To all of our friends and colleagues at RPC, and to all of the earlier generations of employees at RPC, this book is dedicated to you.

We would like to acknowledge and thank Dr. Shelley Rinehart, Linda Horsman, Amy Brown, Tracy Lean, Lise Morin, Steve Holmes, Dr. Diane Botelho, Ann Marie Schramm, and William Schramm for reading drafts, making constructive suggestions and, most importantly, for their support and encouragement.

We are grateful to the following RPC colleagues for providing information and materials for this book: Tracy Lean, Amy Brown, Linda Horsman, Steve Holmes, Matt Ashfield, Jim Carter, Sandi Walker, and Brian Bell. Thanks also to Dr. Ian Potter for providing a history of the Vineland Research and Innovation Centre.

In the preparation of this book our work was also assisted by the staff and collections of the libraries of the University of Saskatchewan and University of New Brunswick, and of the New Brunswick Provincial Archives.

1 RESEARCH AND TECHNOLOGY ORGANIZATIONS[1]

At the beginning of the 20th-century organized research, development, and technological innovation were quite rare in North America, particularly outside of individual companies and government departments.

In Ontario, for example, a number of privately-owned fruit experimental stations were created, beginning in the 1890s, with the aim of determining what varieties of fruits were best suited to the growing conditions in different parts of the province, and to provide guidance to the planters. By the turn of the century, there were fifteen such stations.

In 1905, it was determined that a central, government-owned fruit experimental station should be established in the Niagara district where *"more of the problems confronting the fruit-grower might be worked out than [could] be attempted under the plan of privately-owned stations"* [4]. In 1906, the Ontario government established the Horticultural Experiment Station "Jordan Harbour" on 90 acres of land [4,5]. The Station was part of the Ontario Department of Agriculture, and gradually assumed the work of the fifteen individual stations, which were then discontinued. Beyond simply consolidation, however, the aim of the Station was to carry out [4]:

"more advanced experimental work than was possible on grower-owned and operated 'variety-test' plots ... provide needed expansion from variety testing mainly, to controlled investigations of cultural problems, pollination, soil management, pruning, propagation, pest control, etc.; and the actual breeding of new varieties suited to Ontario conditions and markets. As well there would be similar work with vegetable crops."

---

[1] Parts of this chapter have previously been published as Chapter 1 in Reference [6].

Thus was born possibly the first example of what are now known as research and technology organizations (RTOs) in Canada (Table 1.1). The Station has evolved through several name changes[2] over the decades and is now the Vineland Research and Innovation Centre.

Research and Technology Organization (RTO)

A government-owned corporation/agency or private not-for-profit company that is primarily focused on developing and deploying practical technologies that address commercial marketplace problems or opportunities. RTOs generally span multiple sectors of an economy. RTOs differ from academia, mainstream government, for-profit companies, and even from mainstream not-for-profit companies. They have "public good" missions and are not primarily profit-driven. They work in the public interest but on market-pull issues, generally using a businesslike approach, and frequently with a secondary, socio-environmental agenda. One of the most important functions of an RTO is to help business enterprises access, absorb, adapt, deploy, and exploit new technologies in order to enhance businesses' ability to innovate and therefore their competitiveness and sustainability. Provincially owned RTOs in Canada are sometimes referred to as provincial research councils or Provincial Research Organizations (PROs).

The early 1900s marked what is sometimes referred to as the scientific revolution, as industries of all kinds became increasingly science-based and industrial research organizations began to spring up [6]. For example, the United States' National Research Council (U.S. NRC) was formed in 1916 and seems to have quite quickly become interested in the role of research in helping industries advance (hence the term *"industrial research"*).

In 1928 Maurice Holland, the first head of Engineering and Industrial Research at the U.S. NRC, published what may have been the first description of the steps involved in the *"cycle of research"* leading to technological innovation and industrial growth [7,8]. Holland was an early champion of systematic research, to which contributions are made by numerous researchers, as opposed to the independent, lone-wolf type of

---

[2] In 1966, the Horticultural Experiment Station became Horticultural Research Institute of Ontario (HRIO), in 1997 HRIO was transferred to the University of Guelph (along with the Simcoe Horticultural Research Station and the Muck Crops Research Station, Kettleby), and in 2009 it was spun-out as the not-for-profit Vineland Research and Innovation Centre.

inventor. Systematic research and development (R&D), and the linear model of technological innovation were fairly new concepts at the time. In 1928, research was still a relatively new tool of industry in North America - two of the first U.S. industrial research laboratories had been established by General Electric (in Schenectady, New York in 1900) and Du Pont (in Wilmington, Delaware in 1903). Although Holland's model later turned out to be an oversimplification, the concepts of a development pathway and a systematic approach were critical to the evolution of the modern approach to technological innovation [9].

Although Canada, for its part, had begun earlier with the 1906 establishment of what is now Vineland Research and Innovation, the U.S. NRC provided an example of an organization that could undertake broad, multi-sector industrial research. Leading Canadian industrialists and university presidents had been calling for some kind of national effort, broad in scope, to boost scientific research in Canada and the British government, which had just created a Department of Scientific and Industrial Research (in 1915), had asked that each of the Commonwealth countries do the same in order that they might all collaborate [10].

In 1916, the same year that saw the formation of the U.S. NRC, Canada formed its own research council of national, multi-sector scope: the Honorary Advisory Council for Scientific and Industrial Research. Canada's research council was created by an Act of Parliament in 1917 (Bill 83), and in 1925 it became known under its "short title" as the National Research Council (NRC) [10,11].

The initial objectives of NRC were to promote the application of science to Canada's "*secondary industry[3],*" to advise the federal government on scientific matters, and to "*put science and industry together for the benefit of the people of Canada*" [11]. NRC's first offices were located in the West Block of the Parliament Buildings (in 1917). NRC moved into its first, modest laboratory facilities in 1926, and to its first large, dedicated laboratory building[4] (on Sussex Drive in Ottawa) in 1930-32 [11].

NRC's first major project was studying the briquetting of Saskatchewan lignite to replace imported anthracite coal as a fuel [10,11], a project that the Saskatchewan Research Council (SRC) would take up in 1947. NRC also established a Bureau of Standards in 1924 [10]. From the opening of its first laboratory in 1927, it was determined that NRC would pursue a balance of 'pure' and 'useful' research[5] [11]. Accordingly, it focused on scientific and

---

[3] That is, industries beyond agriculture and natural resources, for which existing federal government departments already had responsibility.

[4] NRC's "*central research institute*," in its original, grand Sussex Drive building, has been referred-to as the "*temple of science*" [10].

[5] Now more commonly referred-to as "discovery" and "applied" research, respectively.

industrial research related to Canada's efforts during the First and Second World Wars with major programs in chemical and biological warfare agents, explosives, radar and sonar, and atomic (fission) energy. Following the Second World War, NRC's focus almost completely shifted to research that was felt to be in the national public interest but not well provided by either the academic or industrial research sectors [12]. This meant establishing linkages with both academia and industry while continuing to focus on a combination of discovery research and applied research initiatives.

Table 1.1 Some Canadian Research and Technology Organizations (RTOs)

Inception	Original Name	Current or Last Name
1906	Horticultural Experiment Station "Jordan Harbour"	Vineland Research and Innovation Centre
1916	Honorary Advisory Council for Scientific and Industrial Research	National Research Council of Canada
1921	Advisory Council of Scientific and Industrial Research of Alberta	InnoTech
1925	Pulp and Paper Research Institute	→ FPInnovations
1928	Ontario Research Foundation	Process Research ORTECH
1930	Research Council of Saskatchewan	Saskatchewan Research Council
1944	British Columbia Research Council	BC Research Corp.
1946	Nova Scotia Research Foundation	InNOVAcorp.
1962	New Brunswick Research and Productivity Council	Research and Productivity Council (RPC)
1963	Manitoba Research Council	Industrial Technology Centre
1966	Petroleum Recovery Research Institute	→ InnoTech
1969	Centre de Recherches Industrielle du Québec	
1975	Forest Engineering Research Institute	→ FPInnovations
1975	Centre for Cold Ocean Resources Engineering	C-CORE
1975	Prairie Agricultural Machinery Institute	PAMI
1979	Forintek Canada	→ FPInnovations
1983	Centre for Frontier Energy Research (C-FER)	C-FER Technologies Inc.
1985	l'Institut national d'optique	INO
1987	PEI Food Technology Centre	Bio | Food | Tech
2007	FPInnovations	
2009	Research and Development Corp.	RDC

*The symbol → denotes the result of a merger or acquisition.

The first provincial research council to be established in Canada was in Alberta (see Appendix 8.9). The province had begun to actively evaluate its mineral resources and their economic development potential, and it was sufficiently encouraged by the results that The Advisory Council of Scientific and Industrial Research of Alberta[6], was established in 1921 *"with a view to ascertaining more definitely the mineral resources ... and the possibilities of their development"* [13,14]. Alberta's research council work was halted between 1933 and 1942 due to the Great Depression but resumed operations in 1942 [15], originally working in co-operation with the University of Alberta and on the university campus, but eventually moving to its own dedicated facilities in 1956 [16].

In 1925, the Pulp and Paper Research Institute of Canada (Paprican) was created as a not-for-profit RTO. As the name implies, it was focused on pulp and paper research, with national-scope applied research, development, and testing programs supported by industry and government members. It was amalgamated into FPInnovations in 2007 (see below, in this section).

The next provincial research council to be created in Canada was the Ontario Research Foundation (ORF, announced in 1927 and established by a provincial Act in 1928) [10,17]. A separate Research Council of Ontario was created in the mid-1940s but was dissolved in 1955 and most of its activities were transferred to ORF [18]. ORF was renamed ORTECH Corp. in 1990 [19] and privatized in 1999. See also Appendix 8.9.

The third provincial research council to be established in Canada was the Research Council of Saskatchewan, under *"The Research Council Act, 1930."* Unfortunately, this coincided with the beginning of the "Great Depression" in Canada in which the prairie provinces were particularly strongly affected (mostly due to the collapse in wheat prices), and which was to last until the beginning of the Second World War in 1939. Although some early work was conducted, related to natural resource development, the prevailing economic conditions made it impossible to conduct active applied research and development programs, and the Act was repealed in 1935 [20,21]. Following the end of the Second World War, attention returned to the mineral resource potential and opportunities to grow and diversify Saskatchewan's economy beyond agriculture. As a result, there was a renewal of interest in coordinating research activity [20] and the Saskatchewan Government re-established the research council under a new name, Saskatchewan Research Council (SRC) in 1947. Under the Act, SRC's primary mission was to *"take under consideration matters pertaining to research and investigation in the field of physical sciences as they affect the economy of the Province of Saskatchewan."* This was broadened in 1978 to *"take under consideration matters pertaining to research, development, design, consultation, innovation and investigation in, and commercialization of, the natural and management sciences, pure and applied, as they affect the welfare of the province"* [22].

---

6 Later renamed Science and Industrial Research Council of Alberta, Research Council of Alberta, Alberta Research Council, Alberta Innovates – Technology Futures (AI-TF), then InnoTech (a subsidiary of Alberta Innovates).

The fourth and fifth provincial research councils to be established in Canada (see Appendix 8.9) were the British Columbia Research Council (incorporated as a not-for-profit society in 1944) [23], and the Nova Scotia Research Foundation (NSRFC, established by a provincial Act in 1946) [24]. Neither would be particularly long-lived. The British Columbia Research Council was privatized as BC Research Corp. in 1988 [25], and NSRFC was amalgamated into InNOVAcorp., a venture capital firm, in 1995 [26].

The next provincial research council to be established was in New Brunswick. "*An Act to Establish a Research and Productivity Council*" was assented on April 13, 1962 (see Figure 2.1) [27,28]. The New Brunswick Research and Productivity Council is now better known via its trade-name: RPC. RPC's story begins in Chapter 2.

The Manitoba Research Council, the seventh provincial research council, was formed in 1963. It was subsequently split up to form the Food Development Centre, in 1978, and the Industrial Technology Centre (ITC) in 1979.

The Petroleum Recovery Research Institute (PRRI) was created as a joint effort of the Canadian petroleum industry and the Alberta Government. Originally based in Calgary, its mission was to focus on conventional oil recovery, particularly enhanced oil recovery (EOR) in Canada. It was incorporated as a not-for-profit research and engineering organization in 1966. In 1974, the name was changed to Petroleum Recovery Institute (PRI). PRI was acquired by ARC in 1999 and later merged into ARC's (now InnoTech's) Energy Division.

The province of Québec formed a research council, Centre de Recherches Industrielle du Québec (CRIQ) in 1969 [27].

In 1975, the Forest Engineering Research Institute of Canada (FERIC) was created as a not-for-profit RTO. Originally based in Pointe-Claire, Québec, it was focused on forest harvesting and transportation, with national-scope applied research, development, and testing programs supported by industry and government members. It was amalgamated into FPInnovations in 2007 (see below).

The Centre for Cold Ocean Resources Engineering (C-CORE) was also created in 1975 as a not-for-profit research consortium of the Canadian petroleum industry, Memorial University, and the Newfoundland and Canadian governments. Based in St. John's, its original mission was to focus on the research and development challenges facing Canada's oil and gas development offshore Newfoundland and Labrador, and other ice-prone regions. The mission was later broadened to include harsh environments generally.

The Prairie Agricultural Machinery Institute (PAMI), is a Humboldt, Saskatchewan-based, not-for-profit, research and technology organization (RTO). PAMI was created in 1975 as a partnership among the agricultural industry and the provinces of Manitoba, Saskatchewan, and Alberta to create a central agency for testing and evaluating farm machinery in Canada.

In 1988, Alberta withdrew from PAMI to independently operate an Alberta Farm Machinery Research Centre (AFMRC, renamed Agricultural Technology Centre, ATC, in 2001) in Lethbridge. PAMI continues to focus on applied research, development, testing, and evaluation related to farm machinery.

In 1979, Forintek Canada Corp. was created as a not-for-profit RTO. Originally based in Vancouver, B.C. (and with a significant presence in Québec as well), it was focused on wood products research, with national-scope applied research, development, and testing programs supported by industry and government members. It was amalgamated into FPInnovations in 2007 (see below).

The Centre for Frontier Energy Research (C-FER) was created in 1983 as a research consortium of the Canadian petroleum industry, the University of Alberta, and the Alberta and Canadian governments. Based in Edmonton, its original mission was to focus on the development of Canada's Arctic and offshore oil and gas resources. The mission was later changed to operational productivity in the conventional and unconventional oil industry, particularly in wellbore production and pipelines. It was re-incorporated as a not-for-profit research and engineering company in 1997. C-FER was acquired by the Alberta Research Council in 1999, and made a subsidiary: C-FER Technologies (1999) Inc.

L'Institut national d'optique, INO (the National Optics Institute) was created on December 13, 1985, at the instigation of the Canadian Association of Physicists, the Natural Sciences and Engineering Research Council of Canada (NSERC), various partners in the Québec City region, and Laval University which, in aggregate and at that point in time, accounted for half of Canada's researchers in optics/photonics. Incorporated as a not-for-profit research and development company, INO remains focused on optics and photonics research for industrial development purposes.

Prince Edward Island (PEI) established the ninth provincial research council in Canada with the PEI Food Technology Centre, which was incorporated in 1987 as a wholly-owned subsidiary of Innovation PEI, a Provincial Crown Corporation. Its name was changed to Bio|Food|Tech in 2011 (but it remains owned by the province).

In 2007, FPInnovations (FPI) was created by amalgamating Forintek, FERIC, and Paprican. FPI maintains major facilities in Montréal, Quebec City, and Vancouver. Closely associated with FPI is the Canadian Wood Fibre Centre (CWFC, created in 2006), which is formally part of the Canadian Forest Service but which delivers work planned by FPI and works co-operatively with the other three divisions of FPI. Some of the aims of the consolidation were to provide greater efficiency, synergies, and strength related to research, development, and testing on forest products sector issues.

Finally, Newfoundland and Labrador created Canada's tenth and most recently formed research council. It was called the Research and Development Council under its own Act in 2009[7], and later renamed Research and Development Corp. (RDC). RDC was short-lived, having been wound-up in 2018.

The foregoing description only covers a selection of RTOs, but it provides some sense of the origins, breadth, and geographic diversification of Canada's regional and national RTOs – not to mention their rather precarious existence.

In reviewing the establishment of the provincial research organizations (PROs) in Canada, Le Roy and Dufour conclude [18] that:

"The fact that in some provinces the establishment of the PROs took place much later than in others was not because of any lesser concern for industrial development. Rather, it was a question of when the provincial authorities became convinced that the industrial development of their province could be stimulated through the agency of a provincially sponsored body specially dedicated to this purpose."

"Over the years in Canada an impressive network of Research Councils has been built up, … and they have, I think, constituted a great and well-recognized Canadian achievement."
Dr. E.W.R. Steacie [29]
President of the National Research Council

---

[7] A Newfoundland Research Council Act was actually passed in 1961 but not implemented - a new Act was passed in 2009 to create what became the Research and Development Corporation (RDC). See Appendix 8.1.

2 RPC: THE BUILDING YEARS, 1962 – 1969

> *"The costs of technological research are great. The rewards*
> *of such research are of much greater significance."*
> Dr. L.W. Shemilt, Chairman,
> RPC's First Annual Report, March 31, 1963

As noted above, the New Brunswick Government created a dedicated research council with *An Act to Establish a Research and Productivity Council*, on April 13, 1962 (Figure 2.1) [27,28].

The government stated in the Act's first reading in the Legislature that the purpose was to *"stimulate and expedite continuing improvements in productive efficiency and expansion in various sectors of New Brunswick's economy"* [30]. The Act itself, set out the government's intended purpose for the New Brunswick Research and Productivity Council[8] (RPC) as follows [28]:

"The objects of the Council are to promote, stimulate and expedite continuing improvement in productive efficiency and expansion in the various sectors of the New Brunswick economy, and without limiting the generality of the foregoing, the Council shall,

(a) assist, encourage and where necessary conduct research, studies and investigations in the field of industrial and scientific technology, particularly those related to
(i) the conservation, development and utilization of the natural resources of the Province,
(ii) the development and utilization of the by-products of any processes involving the treatment or using of the mineral, timber or other resources of the Province, and

---

[8] Hereafter referred to as RPC.

(iii) the improvement and development of industrial materials, products, and techniques; and

(b) co-operate with the National Productivity Council to foster and promote
(i) the development of improved production and distribution methods,
(ii) the development of improved management techniques,
(iii) the maintenance of good human relations in industry,
(iv) the use of training and retraining programmes at all levels of industry,
(v) the extension of industrial research programmes in plants and in industries as a means of achieving greater productivity, and
(vi) the dissemination of technical information; and

(c) co-operate and act in conjunction with other organizations and agencies, public and private, in the implementation of programmes designed to give effect to any of the objects described in clauses (a) and (b)."

Stated more briefly, the purpose of RPC would be to use research, development, and technological services to work with industry to help grow and diversify the New Brunswick economy. In remarks to the Provincial Legislature a year later [31] Premier Robichaud expanded on this, predicting that RPC would become the scientific arm of the government and *"become the foundation for solid and productive scientific research in the province."*

The provincial government established RPC's head office in Fredericton, and appointed its first Council (i.e., Board of Directors[9]). Befitting the role RPC was expected to play, the first twelve Members of Council were both geographically diverse and represented a cross-section of "industry and commerce, organized labour, government, and higher education" [32] (see Figures 2.2 and Appendix 8.2).:

- Dr. L.W. Shemilt (Chairman), Univ. of New Brunswick, Fredericton,
- Mr. F.R. Drummie (Executive Director), Treasury Board, Province of New Brunswick, Fredericton,
- Mr. K.V. Cox, New Brunswick Telephone Company Ltd., Saint John,
- Mr. W.E. Estes, Engineering Consultants Ltd., Saint John,
- Mr. L.W. Flett, North Shore Construction Ltd., Newcastle,
- Mr. H. Fredericks, Hal Fredericks Associates Ltd., Sussex,
- Mr. M. Kenny, Newcastle,
- Mr. L.H. Lajoie, L.H. Lajoie Ltd., Edmundston,

---

[9] At the time the Board was referred to as the 'Council,' and the Members were 'Members of Council.' Beginning with the 1989/90 fiscal year the terminology was changed to 'Board of Directors' and 'Directors,' respectively.

- Rev. L. Lanteigne, Sacred Heart University, Bathurst,
- Mr. H. McCain, McCain Foods Ltd., East Florenceville,
- Dr. D.B.W. Robinson, Bathurst Power and Paper Company, Bathurst,
- Mr. M.A. Savoie, Barrister, Moncton

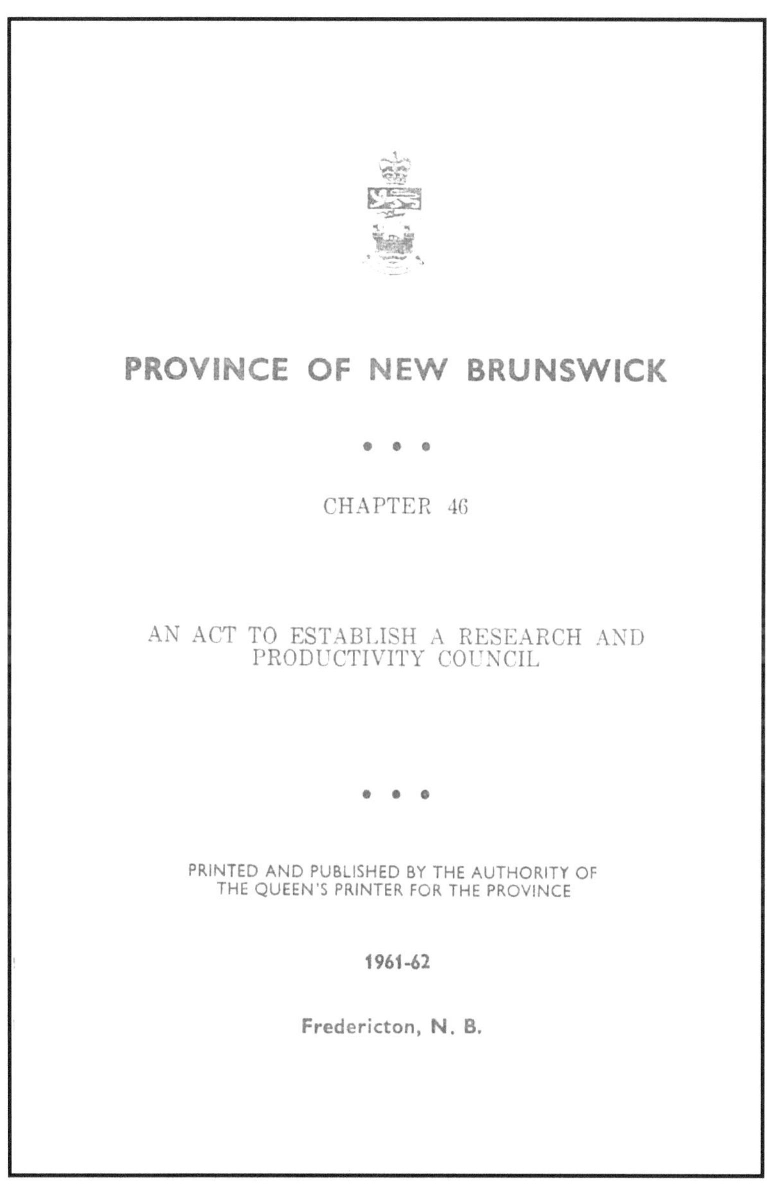

Figure 2.1. The Act to Establish a Research and Productivity Council in New Brunswick, 1962.

The first meeting of RPC's Council was held on September 14, 1962. Some of the initial activities included the drawing up of By-Laws (which were then approved by Provincial Order-in-Council), establishing policies and procedures (such as for inventions and patents), and the staffing and launching of a temporary office at 450 King Street[10] in Fredericton (Figure 2.3).

A top priority for the Council was to find a permanent Executive Director for the new organization. Mr. D.W. Gallagher had already served unofficially as Executive Director for a few months, prior to the appointment of Mr. F.R. Drummie as the first official, Order-in-Council appointed Executive Director. Mr. Drummie's appointment as Executive Director, however, was also only intended to be temporary, and *"an advertising programme to assist in securing a permanent Executive Director of high qualifications was undertaken"* [32].

Figure 2.2. A founding Council member and RPC's first Chairman was Dr. Les W. Shemilt. RPC photo, 1962.

---

[10] The site of present-day King's Place complex.

Figure 2.3. RPC's first office was established in this building on King Street in Fredericton, in March 1963. RPC photo.

Figure 2.4. One of RPC's first pieces of equipment: a teletypewriter, telex #014-46115. RPC photo, 1963.

RPC released its first Annual Report (Figure 2.5) six months after the passage of the Act, summarizing some of the first steps taken by the new organization to enable *"its two-fold role of scientific and technological research activity and of contribution to the many aspects of increased industrial productivity"* [32].

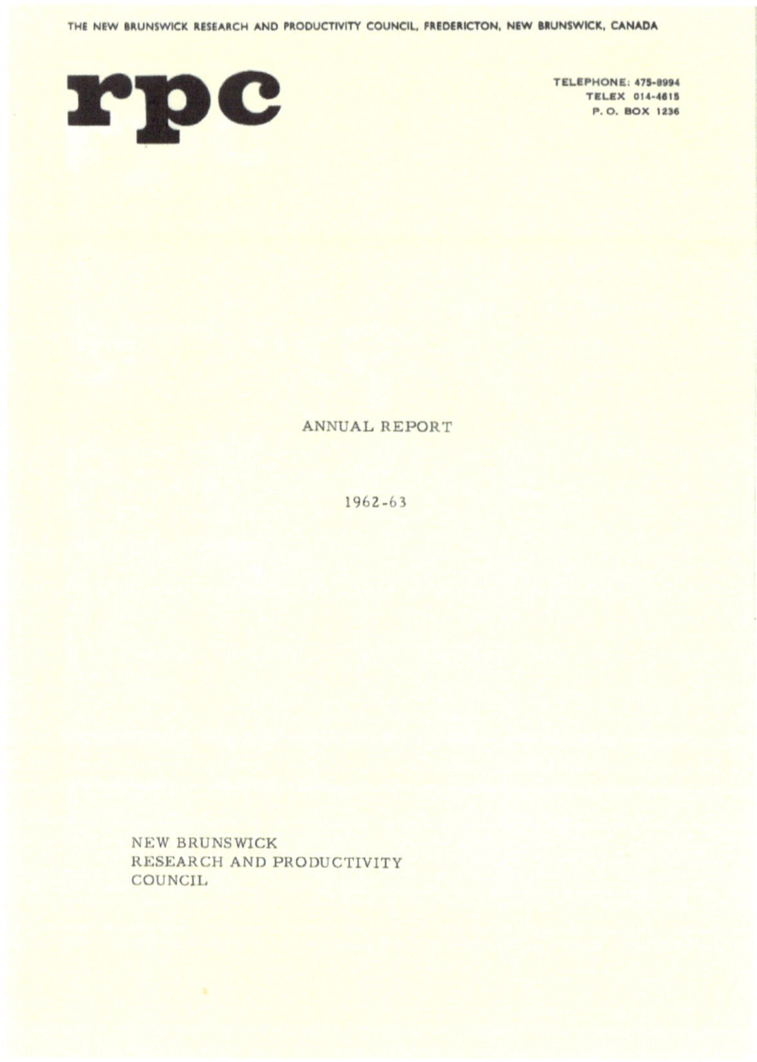

Figure 2.5. RPC's first Annual Report (1962-63), showing its first company logo.

Among the early decisions of the Council were the creation of several Standing Committees. The Research Committee was to focus on *"utilization of the Province's research facilities for a programme relevant to resource exploration and exploitation."* In other words, RPC would begin to fund specific applied research projects to be conducted at the province's universities[11]. The Productivity Committee was to focus on *"developing a programme that will assist in productive efficiency."* The Finance Committee was to focus on *"budgetary matters, and the development of financial support additional to that supplied by government."* A Public Relations Committee was also established.

The first Annual Report makes it clear that, from the very beginning, RPC anticipated conducting direct research activities at some point in time, and not simply funding work to be done by others. In the Chairman's Report, Dr. Shemilt noted [32] that:

"As the scientific and research arm of the government, the Council will and must develop facilities suitable for fulfilling its role. To an increasing extent it hopes to be a scientific, research, and technical service arm for the industry of the Province."

As such, people and equipment would also be needed, and consistent with this, the Finance Committee was to look into the acquisition of additional revenues to bolster the direct investment from the province.

Given the broad- and long-term nature of the objectives set out for RPC under its Act, the first Council formulated some guiding principles [32]:

o   *"Research provides the technological basis for expansion and new business and erects a defense against technological obsolescence that otherwise results in business loss… the resources we possess are not enough in themselves… it takes work and effort to translate them into the products and services we require,"* therefore,

o   RPC would need to *"select wisely areas of concern where significant progress can be made,"* while recognizing that *"results, even in well-defined problems of limited nature, do take considerable time, and that research is by its very nature unpredictable in terms of time and even eventual success".*

With these words, the Council outlined RPC's first program of work, which comprised two broad themes, a Research Program and a Productivity Program.

---

[11] These university grants started in 1962, and were wound-down to zero by 1967/68.

The **Research Program** involved the awarding of grants for specific university projects (Figure 2.6). The first year's grants, totaling $35,000, were announced on November 10, 1962 [33]. The funds were used to support direct operating costs such as student research assistants, supplies, and equipment, while the university faculty themselves, donated their time. The first main applied research projects were aimed at the province's diverse natural resources, including fisheries, and manufacturing. Noteworthy among the first projects were [32]:

- o Conducting a detailed gravity survey of the province, so that gravity anomalies calculated from survey data could be used to aid mineral and oil exploration[12], and
- o Building a facility for the design and development of electronic equipment such as could provide the basis for an electronics manufacturing industry in the province.

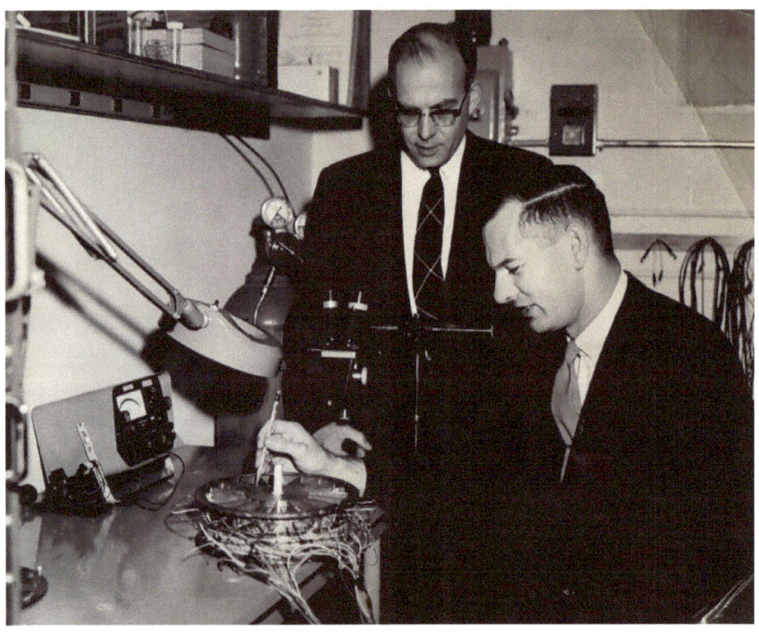

Figure 2.6. RPC Chairman Dr. Les Shemilt (standing) advises University of New Brunswick (UNB) Electrical Engineering Professor R.N. Scott that he has been awarded one of RPC's first applied research grants. RPC photo, 1962.

---

[12] Later published as Konecny, G., "Gravity Survey of the Province of New Brunswick, Part 1," Technical Report No. 3, University of New Brunswick, Fredericton, 1967.

The **Productivity Program** involved three areas of activity. In the first area, personnel from industry and government were recruited to participate in training sessions in *"modern management and industrial production techniques."*

The first "management training program" announcement was made in Fredericton on April 13, 1963 [34] (see Figure 2.7), and additional offerings were made in other cities and also in conjunction with the Nova Scotia Technical College. These were well received, and according to media reporting [35] *"Recent evaluations made by firms who have had their employees participate in the school is very encouraging and all have been most enthusiastic with the results obtained. Investment of several hundred dollars have proven to be savings in the four and five-figure bracket."* Encouraged, RPC continued and expanded its training program (Figure 2.8). By May 1965, 300 people from 120 companies had participated in these sessions [1].

In the second area, RPC partnered with the National Research Council to develop and deliver technical information and field engineering services to industry. In the third area, it was planned to develop and deliver *"short courses in various aspects of management training where such needs are not being met by other agencies."*

Figure 2.7. RPC advertisement for its first Management Training Program April 23, 1963 [36].

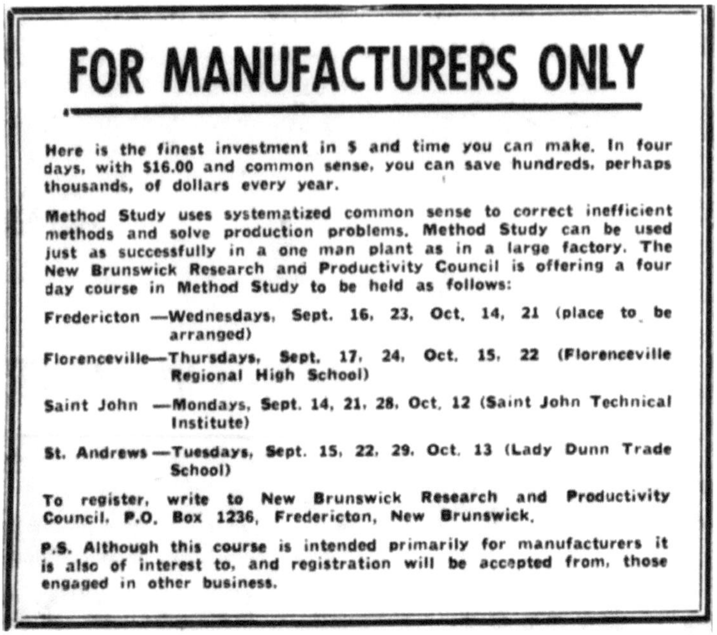

FOR MANUFACTURERS ONLY

Here is the finest investment in $ and time you can make. In four days, with $16.00 and common sense, you can save hundreds, perhaps thousands, of dollars every year.

Method Study uses systematized common sense to correct inefficient methods and solve production problems. Method Study can be used just as successfully in a one man plant as in a large factory. The New Brunswick Research and Productivity Council is offering a four day course in Method Study to be held as follows:

Fredericton —Wednesdays, Sept. 16, 23, Oct. 14, 21 (place to be arranged)

Florenceville—Thursdays, Sept. 17, 24, Oct. 15, 22 (Florenceville Regional High School)

Saint John —Mondays, Sept. 14, 21, 28, Oct. 12 (Saint John Technical Institute)

St. Andrews —Tuesdays, Sept. 15, 22, 29, Oct. 13 (Lady Dunn Trade School)

To register, write to New Brunswick Research and Productivity Council, P.O. Box 1236, Fredericton, New Brunswick.

P.S. Although this course is intended primarily for manufacturers it is also of interest to, and registration will be accepted from, those engaged in other business.

Figure 2.8. RPC advertisement for Method Study Training Sept. 5, 1964 [37].

RPC's first three employees, hired early in 1963, were Miss J. Nixon (Office Secretary), H.J Fullerton (Research Engineer), and P.B Aitken (Technical Information Officer). Total revenue for RPC's first year of operation comprised a provincial grant of $50,000, and $25,796 was paid out in research and special equipment grants to universities [32].

In 1963, Dr. Claude Bursill was hired as RPC's first "permanent" Executive Director [38]. He served in this capacity until September 1983 [39] (see Figure 2.9 and Appendix 8.3). The creation of RPC and the hiring of Dr. Bursill were both viewed favourably in the media. In an editorial titled "Theirs is a Challenging Task," for example, the *Moncton Daily Times* praised these inaugural steps [40] saying, in part:

> "*The council seems likely to be able to provide most beneficial services to industry. By assisting industry to take advantage of recent technical developments, and by undertaking research related to the resources of New Brunswick, the council should be able to stimulate greatly industrial expansion within this province... There is a very great deal of work to be done before New Brunswick is even near to catching up with other parts of the country in industrialization. But sure steps forward are being taken and not the least important is the establishment of the Research and Productivity Council with its excellent membership.*"

Figure 2.9. RPC's first permanent Executive Director, Dr. Claude Bursill, took office in 1963 and served until 1983. RPC photo.

Dr. Bursill had a clear sense of mission for RPC and all of the other provincial research councils, which he articulated in a January 5, 1965 letter he sent to the editor of *The Economist* on "Scientific Policy [41]. An edited version appeared in the January 16, 1965 issue [42].

In his letter, he addressed:

"the relationship between industry and research effort in regions [around the world] where the greater part of industry is not aware of the advantages of research, has not the wherewithal to undertake it, or is dependent on foreign research. Similar conditions exist in parts of Canada, and several Canadian provinces have devised a solution which should have general application to under-developed regions...

The Canadian device is a regional technological research facility, a research council independent of the Civil Service, supported partly by provincial funds and federal grants, and partly by repayments from industry. This practical arrangement permits the growth of regional institutions staffed by scientists and engineers with industrial experience, who are devoted to scientific and technological research for industry. They are multi-discipline institutions, each varying in its policy to suit regional conditions. They include a group of engineers who visit industry constantly, offer confidential advice and assistance in industrial engineering, a technical information service, and give courses of lectures on management techniques. In this way, a demand for improved technology and research is built up in local industry, and there is continual feed-back to the scientists. The scientists and engineers have the pleasure of being committed to the advancing of local industry, with which they are familiar in detail, and measuring the return on their effort. The work of these Canadian provincial research councils is quite neatly woven into the more basic, or more expensive, effort of universities and federal agencies, and into the economic planning and industrial development of the provincial governments."

Among Dr. Bursill's early priorities were to inventory the research already underway in the province, the recruiting of additional professional staff, and the planning for a dedicated headquarters and laboratory, and pilot plant facilities. As these began to take shape, it became possible to increasingly focus on helping companies improve productivity and lay the foundations for longer-term research and development aimed at new products, processes, and resources [43].

One of the most important of Dr. Bursill's early priorities was the research inventory, as it would identify the first opportunity areas for the new organization. Accordingly, RPC organized a two-day Research Inventory Conference in November 1963 (Figure 2.10). Held in Fredericton, it was opened with remarks from Premier Robichaud [44] and well attended by 158 representatives from industry, government, and academia.

Based on the conference, RPC published its first report: "Catalog of Research Resources and Activities, New Brunswick, 1963" [45]. The catalog spanned natural resource-related and industry areas, such as agriculture, fisheries, forestry, water, utility power, mines and minerals, among others.

This initiative was well received in the media, earning editorial comment by the *Fredericton Daily Gleaner* [46]:

"By promoting this [research inventory] conference the Research and Productivity Council has shown New Brunswick that it is on the job."

Figure 2.10. Research Inventory Conference organized by RPC, Nov. 21-22, 1963. Standing in the photo are (from left to right) Dr. Bursill, RPC Board Member Dr. D.B.W. Robinson, Hon. Louis J. Robichaud (Premier of New Brunswick), Dr. Shemilt, and two unidentified participants. RPC photo.

The recruiting of additional staff was well received in the province. In an editorial titled "To Assist Forward March of Industry in N.B.," for example, the *Moncton Daily Times* praised RPC's rapid expansion [47] saying, in part:

"These appointments, it can be rightly said, furnish significant evidence that the Research and Productivity Council is making sound moves which, with able, energetic and enterprising direction, give promise of a definite forward march taking place not only towards the greater but the more modernized industrialization of New Brunswick. The people of the province will focus an avid interest on the council's enlightened program to advance our industrial growth and which, in turn, would enhance the value of our economic structure and the general prosperity."

With the hiring of additional staff, Dr. Bursill organized RPC into three divisions: Natural Resources Research, Industrial Research, and Industrial Services. Although the structure changed in subsequent years, this list shows what were thought at the time to be the most promising areas for RPC:

- o **Natural Resources Research,**
  - o Mineralogy and Geology,
  - o Hydrology/Hydraulics,
  - o Biology

o **Industrial Research**
  o Electrical/Mechanical Engineering,
  o Food Technology,
  o Analytical Services,
  o Instrumentation,
  o Chemical engineering,
  o Highway Research
o **Industrial Services**
  o Technical Information Service,
  o Industrial Engineering,
  o Work Study,
  o Business Administration,
  o Lectures, Seminars, etc., and
  o Forest Productivity

In developing and implementing the above structure, one of the unstated but significant changes in approach was the winding-down of the practice of awarding grants for specific university projects. The last such payments were made in 1967. In the 1966/67 Annual Report [48], Dr. Shemilt referred to *"a careful consolidation of effort towards the assistance of provincial industry, an increased technological capacity."* Dr. Bursill referred to the priority of enlarging RPC's industrial engineering capability in the 1967/68 Annual Report [49].

It seems clear that RPC's leaders thought it best to devote the Council's financial resources on the employment and equipping of permanent, qualified staff that would focus on a full-time program of practical scientific and engineering research, testing, and development for industry. RPC was not unique in this regard. A decade earlier, the Saskatchewan Research Council had wound-up its university granting program for the same reasons [6].

As the staffing-up progressed, Dr. Shemilt asserted to the media in May 1965 ([50]): *"Let me assure you that we already have a group of which any scientific or technical organization, at any level provincial, national or international- would be proud."* As far as the laboratory facilities were concerned, Dr. Shemilt explained [51] that they were needed to provide:

"direct assistance to New Brunswick industry, particularly in those cases where research and advanced workshop facilities are beyond the [industries'] capability… in areas of special concern to the province, such as in the mineral industry and in the food industry".

While planning for the first permanent building was underway, three rooms at the 450 King Street premises were adapted to permit laboratory and workshop activities [52]. Meanwhile, working with the provincial government, RPC arranged for six acres of land to become part of the UNB campus, which would be made available for the new building [38]. Planning was initiated (1963/64) for an initial laboratory block of 10,500 sq. ft. The 'sod-turning' was in May 1964 (Figure 2.11) and the target occupation date was Fall 1964, although it actually took until Spring 1965.

Figure 2.11. Sod turning ceremony, May 1964, to launch construction of RPC's 10,000 sq. ft. building in Fredericton. From left to right are: Dr. Shemilt, Colin B. McKay (UNB President), Dr. Bursill, M.S. Chan (Head, RPC Food Group), and Premier Robichaud. RPC photo.

RPC's first permanent building was opened on May 21, 1965 [43], housing a permanent laboratory, workshop, drawing office, and office space. Premier Robichaud presided over the grand opening (Figure 2.13) and used an oxyacetylene torch to cut a stainless-steel ribbon. Stainless steel had been chosen as the ribbon material to symbolize modern technology, and the idea of the torch was to symbolize industrial processes [1].

"An important part of the task…" said the Premier, *"is to exercise careful judgement in choosing fields of research which have a good probability of increasing productivity in existing or new industry… [and that] the government expects assistance to New Brunswick industry in the application of available technology, and the development of new techniques of direct application in the province. Such service will, without doubt, contribute significantly to increased production and productivity and a better standard of living for us all. Expansion, change, new developments – these are becoming the order of the day in New Brunswick"* [1].

Although the RPC staff were reported to be pleased with the new building, even before its official opening Dr. Bursill noted [53] that he *"was not disappointed in his expectation that there would be an immediate outcry for twice as much space as everyone [had]."*

Accordingly, plans for expansion and additions were begun almost immediately.

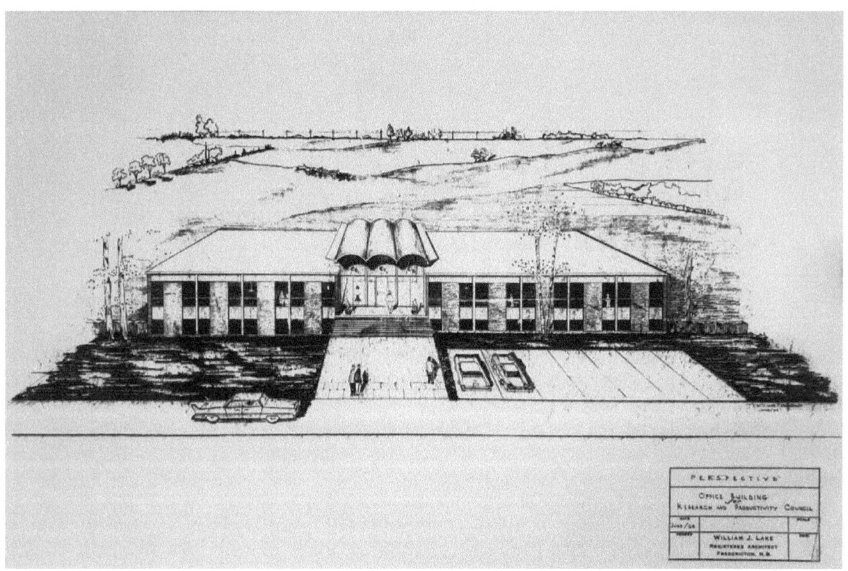

Figure 2.12. Architectural drawing for RPC's first permanent building. William J. Lake, architect, June 1964.

Figure 2.13. Premier Robichaud speaking at the official opening of RPC's first permanent building on May 21, 1965 (upper) and cutting a stainless-steel ribbon with an oxyacetylene torch (lower, wearing lab coat and goggles). Looking on are Dr. Shemilt (upper and lower) and Philip B. Aitken (RPC Technical Information Officer, lower). RPC photo.

Figure 2.14. Photograph of RPC's first permanent building at its official opening on May 21, 1965. RPC photo.

Figure 2.15. Dr. Reginald Gallop, Head of RPC's Food Science Section, demonstrating an automatic titrator at the official opening of RPC's new building on May 21, 1965 [1].

A small (3,600 sq. ft.) high-head pilot-plant was added in March 1966 (Building #2), and planning was underway to expand into an adjacent 8,000 sq. ft. laboratory block (Building #3, previously the Veterinary Laboratory, Department of Agriculture). The former Veterinary Laboratory was acquired, renovated, and occupied by RPC's Food Sciences Department in April 1967 [54] (see Figure 2.16). Meanwhile, planning continued for the construction of an additional block of laboratories and pilot-plant space [55] (Building #4, with another 45,000 sq. ft., which was opened in 1968).

RPC's buildings would be incomplete without instruments and equipment, of course, and over the next few years a variety of instruments were acquired, including an atomic-absorption spectrometer for dissolved metals analyses, an X-ray diffractometer for mineral identifications, and an Instron material-strength-testing machine. In early 1968, a special equipment grant was received from the Atlantic Development Board, half of which was devoted to the establishment of a metrological laboratory for precision testing and maintenance of quality control of production standards [56].

The new building #4's planning and construction were undertaken in parallel with two years' worth of complicated negotiations with UNB and the province regarding the ownership of the land. By spring 1968, however RPC had reached an agreement with UNB by which nearly 10 acres of land was transferred to the university in return for a renewable lease[13] for 99 years. The new building was completed and occupied in April 1968, comprising 45,000 sq. ft. of engineering and mineralogy laboratories, stores, management engineering offices, library, and conference room. This gave an overall total of 65,000 sq. ft., and provided the library with a stack capacity of 30,000 volumes (Figure 2.17).

The grand opening of building #4, on June 5, 1968, included such dignitaries as the Lieutenant-Governor, the Minister Responsible for RPC, several other Cabinet Members, RPC's Council Members, community leaders, and even included the presidents of two of RPC's sister organizations: the Saskatchewan Research Council and the Nova Scotia Research Foundation. The opening remarks by Dr. Shemilt were broadcast live in a 22-minute CBC Radio Special [57] and included the following:

"Our Act gave us structure. Our government has given us strength by its support. This support has been not only financial, I might say, but the government too has been our advocate in representations to the federal government's agency, the Atlantic Development Board, and in that board there has been equal and committed concern for the policy that applied science, technology, innovation, [and] new knowledge must play an important role in the improvement of productivity and the achievement of economic growth..."

---

[13] The lease was signed in November, 1968.

Dr. Shemilt was also quick to emphasize the importance of RPC's employees, saying that, although the new laboratories gave RPC:

"a significant scientific and technological capability… Nonetheless, means to experiment is not just buildings and sophisticated instruments. More than anything else it requires conviction and commitment, the brains and the hearts, of people…" [57].

His conclusion also bears recording here:

"Our purpose is really simple. To provide on the front of modern science and technology, assistance to the economic and industrial life of the province. We're here really, dedicated to this, to working with and for industry, a purpose which is particularly cogent in terms of having such facilities immediately available, with the ability to select areas of pertinent concern" [57].

Finally, Dr. Shemilt later noted, in the 1968/69 Annual Report [58], that this was *"the first year in the Council's existence when, for the whole period, adequate laboratory space and facilities have been available."*

In 1963 RPC launched a foundational initiative to gather together the results of all prior prospecting and geological work in the province. Led by Dr. Donovan Abbott, the Head of the Mineralogical Group, the aim was to publish an authoritative source-book of information that would encourage and guide further prospecting and mineral development in the province [38]. Parts A (Industrial and Non-Metallic) and C (Bibliography of New Brunswick Geology) were completed and published in 1963/64, while Parts B (Metallic) and D (The Extent of Geological, Geophysical and Geochemical Investigation in New Brunswick) were completed and published in 1964/65 (see Figure 2.18).

Abbott, D., "The Occurrence of Economic Minerals, Rocks and Fuels in New Brunswick," Parts A-D, Record, New Brunswick Research and Productivity Council, Fredericton, 1965.

Following the publication of this source-book, it was decided [59] that:

"having taken stock of the geological situation in New Brunswick… we could usefully busy ourselves by instituting an investigation into the composition of the ore minerals. This will lay emphasis on the minor element contents and the exploitation of these metals."

This kind of work would become an enduring pillar of RPC's work (Figure 2.18 shows a 1968 example) to the present day.

Figure 2.16. RPC Laboratories, *circa* 1968. RPC photos.

Figure 2.17. RPC Buildings #1 through #4, *circa* 1969. RPC photo.

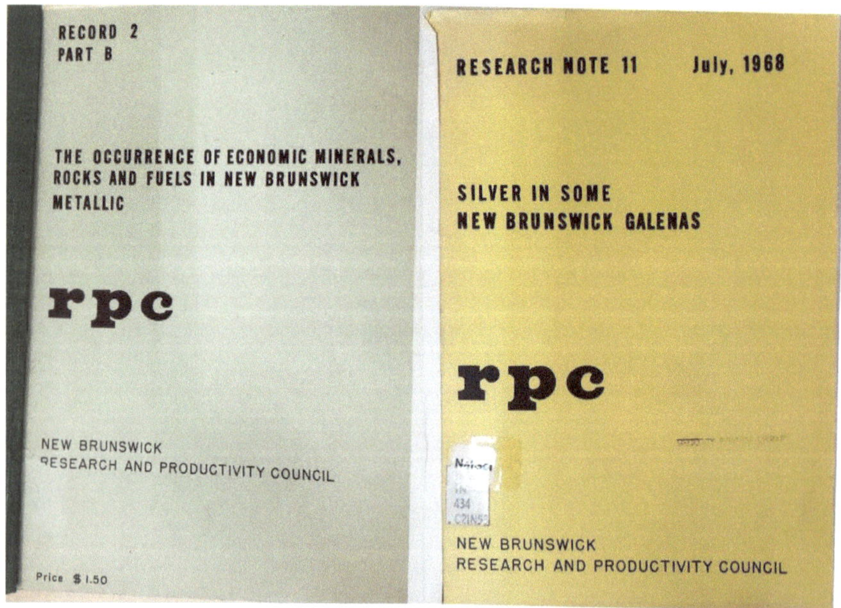

Figure 2.18. Two early examples of RPC publications, one of four parts to the mineral occurrence sourcebook (upper left) and one of the mineral composition treatises (lower right). From the University of Saskatchewan's library collection.

Numerous other publications were to follow the Abbott report, and RPC's first public listing of external publications arising from its work comprised a list of 6 reports, papers, theses, and technical presentations spanning 1962 - 1964 [38]. This practice, of public listing, was continued in subsequent years.

Also in 1963, RPC assumed regional responsibility for NRC's Technical Information Service (TIS), a program aimed at assisting industry by making appropriate research information available to it. Philip B. Aitken had been hired as Technical Information Officer (one of RPC's first three employees), and RPC started giving lectures to industry on management techniques, including topics such as inventory control and critical path methods [38]. Such activities not only enabled RPC to engage with and help industry, small companies, and the public across New Brunswick, but the funding assistance provided by NRC most likely marked the first external revenue to RPC beyond its core operational grant from the province. By 1965 RPC had hired Dr. Dorothy M. Farmer to manage the expanding Library and Information Service [55]. RPC's TIS also marked the first substantial collaboration with NRC, thus beginning a collaborative relationship that was to continue to the present day.

In late 1964, at NRC's request, RPC began to offer TIS services to Prince Edward Island [60], this extension seems to have continued in PEI until about 1981. RPC continued to deliver the TIS program in New Brunswick until 1981, when these services were consolidated under the new NRC Industrial Research Assistance Program (IRAP), which NRC contracted RPC to deliver throughout New Brunswick[14] (see Chapter 3). The TIS program, and RPC's training programs, were not destined to be RPC's only engagements with industry, however, as from as early as 1964 Dr. Bursill and RPC were clearly planning to take on specific industrial engineering jobs, i.e. contracts [43].

In order to support the core Research, Productivity, and TIS programs, a library and library services were launched and quickly built up. The *Bulletin*, an informational, bimonthly periodical for RPC clients was launched in 1964 [52] and maintained (at least) through 1967. RPC also started publishing a monthly journal, *Food Science Abstracts,* in 1965 (Figure 2.19). This journal was published continuously until the end of 1969, at which time it was phased-out in favour of the international journal *Food Science and Technology Abstracts* [61], to which RPC continued to be a contributor.

It wasn't long before the library holdings exceeded the available space, and the library moved to expanded space in the new Building #4 in May 1968 (Figure 2.20).

---

[14] Until 2000 through 2002, when NRC transitioned to delivering this service directly, with its own staff.

Figure 2.19. The first issue of RPC's journal *Food Science Abstracts* (November 1965). Note that the title seems to have been changed after printing. From the University of Saskatchewan's library collection.

Figure 2.20. The new library *circa* 1968. RPC photo.

With its initial programs well underway, RPC next began to look at leveraging its relationships with local universities, and to developing relationships with other research organizations in Canada. In 1964, for example, RPC helped form an Atlantic Provinces Institute of Forestry - jointly with UNB and other interested agencies. This was aimed at encouraging a higher productivity of utilization of forest resources, and was to be partly staffed by RPC, which would make its own laboratories and equipment available [38].

A good working relationship was already in place with the National Research Council of Canada (NRC), of course, whose TIS program provided opportunities to meet with the research organizations that delivered the TIS program in other provinces. Examples are a meeting of RPC, NRC, Ontario Research Foundation, and possibly others at Horticultural Experiment Station ('Vineland Station') in August 1965 [62] and another involving RPC, NRC, BC Research and possibly others at the Saskatchewan Research Council in August 1966 [63]. Dr. Bursill noted in particular, *"the very good relationships that exist between the provincial research councils"* [64,65]. In 1969, RPC was among the founders of the first association of industrial research councils in Canada (NIRAC, see Appendix 8.10). Such relationships, especially those with NRC and the other provincial research organizations, would vary in scope and intensity from year to year but would ultimately stand the test of time and endure to the present day.

From almost the earliest days of his tenure, Dr. Bursill expressed a desire to demonstrate the impacts for RPC's work. In his first confidential report to the Council [52], in March 1964, he wrote:

"The work carried out for industry and government is often confidential and cannot be reported in detail… It is believed, nevertheless, that somehow or other it will prove to be possible to devise a way of demonstrating and, more or less, measuring, the effectiveness and usefulness of the council's organization and staff from time to time…"

He need not have been overly concerned, however. As will be seen in subsequent chapters, RPC would soon earn the greatest measure of effectiveness and usefulness possible: that of an ever-increasing community of committed industrial clients.

"Management science and application of technology are the two most powerful weapons of vigorous and dynamic companies… Companies with the foresight and courage to thrust forward will reap rich rewards. Yesterday's innovation is today's routine and tomorrow's obsolescence. We must go forward using our skills and abilities to the very maximum effect."

Dr. C. Bursill, Executive Director,
"RPC Motivates," 1968 [72]

One of the first industrial projects involved working with a local inventor, Mr. D. McLeod, to develop a simple mechanical device to pull logs through the woods. It was thought that this invention would, if commercially successful, generate a small royalty stream for RPC [52]. The invention appears to have been patented in the U.S. in 1975 [66], but it is not clear whether the product was commercially successful.

Another significant, early industrial, project involved developing improved trawler winch and winch-control designs for Atlantic Fisheries. This project led to other similar projects and catalyzed the creation of RPC's hydraulics laboratory.

An example of the kind of major equipment that RPC had to acquire for its industrial work is the Acton Electron Microprobe[15], acquired in July 1967 [67] (Figure 2.21).

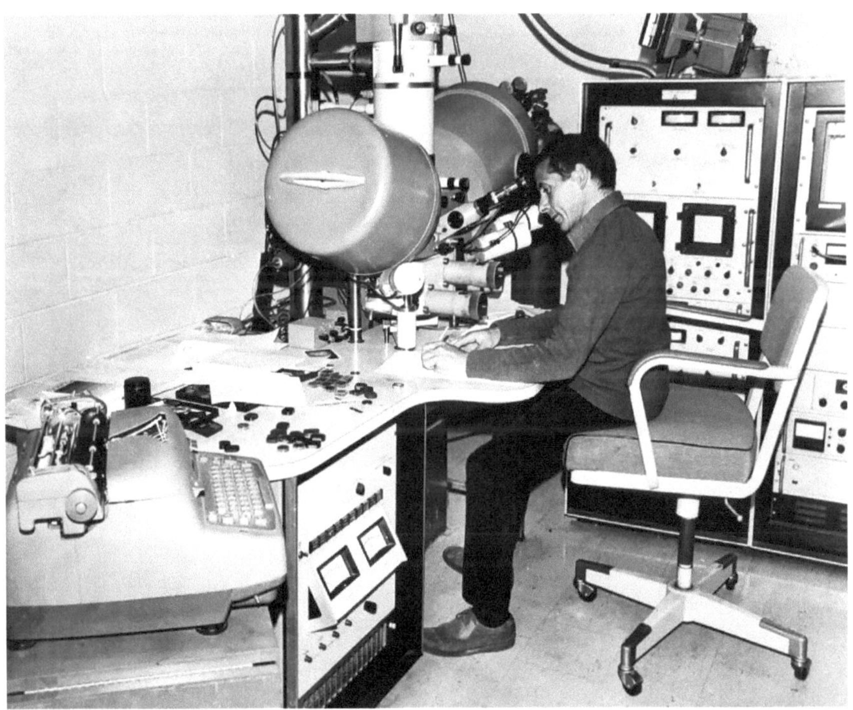

Figure 2.21. Dr. John Sutherland using RPC's state-of-the-art Acton Lab electron microprobe, circa 1968. RPC photo.

---

[15] Probably an Acton model MS64 (the North American version of the Cameca MS46).

RPC even tried making movies to illustrate modern production techniques, and opportunities for productivity improvements, to local industry. This caught the attention of the media as well. An article in the *Fredericton Daily Gleaner* [68] concluded, tongue-in-cheek, that: *"It may not be in the same class as MGM or Warner Brothers, or even the National Film Board, but New Brunswick has its own movie making operation as well."*

Another example of humour in RPC's world was provided by Dr. Bursill himself when, in a July 1969 note on the difficulties being encountered in the development of a mechanical potato-shape-cutting machine, he wrote that the troubles were *"owing to the unfairly anisotropic compressibility of these poorly-designed vegetables"* [69]. Little did he know that RPC would be involved in the development of successive generations of such machines for many decades to come!

As Dr. Bursill explained in RPC's 1967/68 Annual Report [50]:

"We aimed, particularly, at increasing our contract income as one measure of our usefulness and acceptability to provincial industry, at increasing the range of smaller services to help quality control, standardization and product diversification and at a limited number of larger jobs designed either to add greater value to converting raw materials currently in use or to help to introduce new technology into the province."

RPC had begun to advertise its contract services in 1966/67 (Figure 2.22), and the following year (1967/68) had secured and completed some 170 contracts for industry [50].

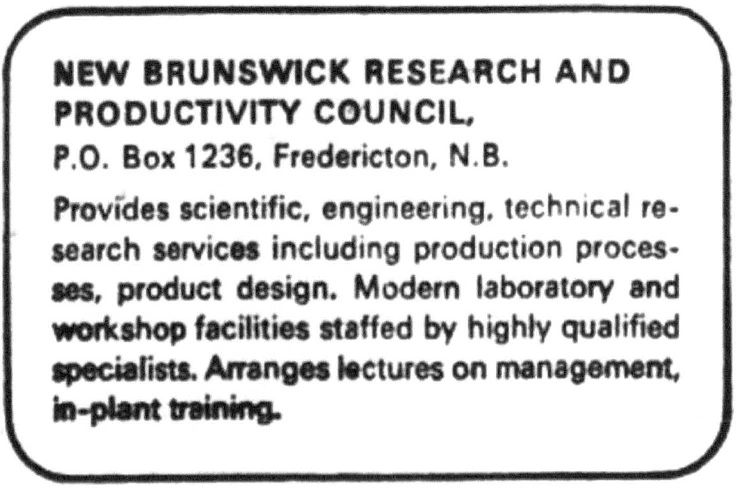

Figure 2.22. Advertising RPC and its services in 1966 [70].

Armed with a growing track-record and client list for contract work, RPC began setting growth targets for its contract revenue, beginning with a target of 22.5% of "revenue from earned operating income" (i.e., revenue other than the core operational grant from the province) in 1967/68 [50]. Contract revenue continued to grow in 1968/69 (in which RPC achieved 26.4% [58]), leading to the establishment of a goal of 30% for 1969/70.

As the industrial work increased, the protection of intellectual property – that of the clients and also that of RPC itself – became important. By 1966 RPC had 3 patent applications underway: one related to a client's design that was modified and improved by RPC, and two originated within RPC [55]. By March 1968, a total of six patent applications were underway. Appendix 8.9 shows how the number of patents listing RPC-employee inventors mushroomed over subsequent years.

In 1968/69 the new machine shop was set up by the Engineering Services Department, with some 27 machines (Figure 2.23), enabling work to commence on projects such as the design of vegetable preparation machinery [58]. Meanwhile, the Electronic Engineering and Instrumentation Department launched new services including product testing, electrical standards, and electrical calibration (Figure 2.24).

Figure 2.23. The machine shop *circa* 1969. RPC photo.

Figure 2.24. RPC's Lyndon Thomas in the electronics shop, 1969. RPC photo.

In 1968 a "homemakers' panel" was created to help with the assessment of new and improved food products [50]. For example, in 1969 the Food Science Department worked on the processing and storage of smoked and jellied eel fillets, which included testing different methods of smoking and different kinds of jellies. According to Dr. Gordon Brown, head of the Chemistry and Food Science Department, *"Eels [had] almost totally been unutilized by provincial fishermen [and] despite the horrible thoughts the slippery creatures bring to mind they are actually a delicious treat, pickled or smoked"* [71]. Based on the lab testing, fifteen hundred pounds of eels were received for preparation of a large batch of two of the products and for frozen eel storage tests. Of the canned varieties, 400 cans of each of the smoked and jellied products were prepared (Figures 2.25 – 2.27). This work was conducted for the Department of Fisheries, and in cooperation with companies including Connors Bros., Paturel Ltd., and W.R. Grace Co.

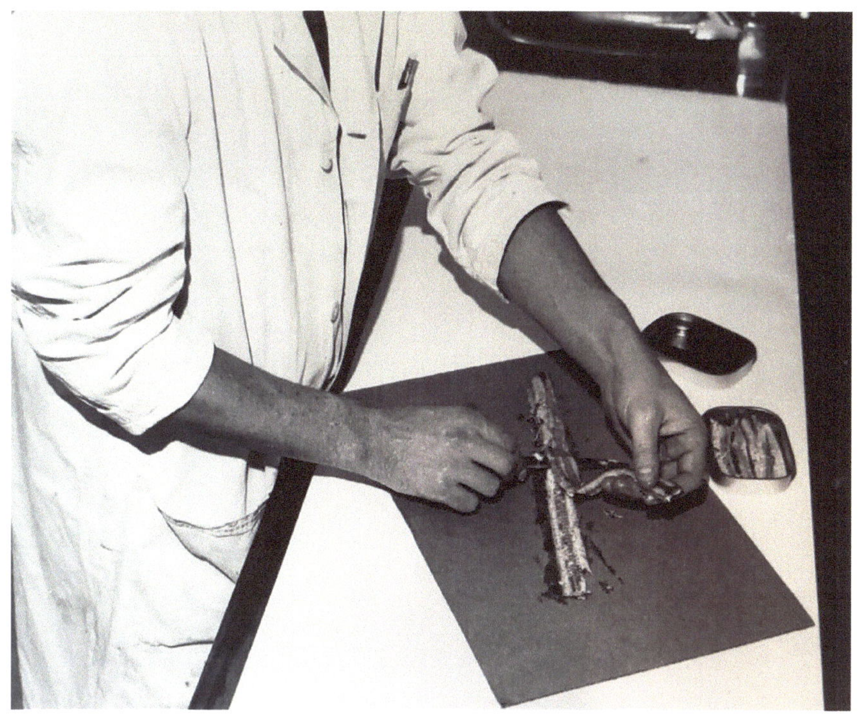

Figure 2.25. Jim Bliss filleting smoked eel, December 1969. RPC photo.

Figure 2.26. RPC smoked eel packs in jelly (left) and soybean oil (right), December 1969. RPC photo.

Figure 2.27. RPC's tasting panel members Charlotte Geldart and Linda Thomson sampling the RPC canned smoked eel products, with Jim Bliss in the background. RPC photo, December 1969.

By this time, RPC had not only conducted hundreds of projects for industry, the impacts of RPC's work on the operations of the companies concerned was becoming readily apparent. Some examples of this were highlighted in a 1968 RPC brochure [72]:

o **Food Processing.** *"One of Canada's latest multi-million-dollar food processing plants is now in its final stages of assessment. From conception to preliminary design the management engineering team undertook all phases of the project,"*

o **Dairy Processing.** *"Work sampling techniques highlighted a gross under-utilization of machine capacity. In addition, introduction of a planned maintenance scheme enable[d] a $15,000 annual savings to be made,"*

o **Fish processing.** *"An annual saving of $25,000 resulted from the development and introduction of a method enabling certain normally rejected fish to be successfully processed for animal feed,"*

o **Clothing Manufacturer.** *"One of Canada's largest apparel and textile industries contracted [RPC] to make an overall corporate review, to solve the low productivity. Engineered work standards and a modern incentive scheme were developed to motivate the workers and increase profits,"* o ***Sorting of Sawmill Lumber.*** *"A survey and report on the most efficient methods of sorting lumber of different dimensions and quality was prepared… working with a major lumber company… [RPC solved] the general problem of optimising the number of service functions in a multi-item service structure. where each function services several types of items but only one item at a time,"*

o ***Machinery Manufacturer.*** *"A manufacturer of agricultural machinery retained RPC in order to overcome the high rate of mechanical failure inherent in the product. The machines were redesigned with the result that the cause of mechanical failure was obviated and the costs of manufacture drastically reduced, allowing the manufacturer to consider world export,"* and

o ***Candy Manufacturer.*** *"A complete appraisal of a major candy manufacturing company was undertaken in order to prepare the company for expansion on a coast-to-coast basis and to lower the costs of production which resulted in proposed initial savings of approximately $50.000.00 per annum."*

Figure 2.28. RPC's Sandra Nearin in one of the early laboratories. RPC photo, 1969.

Figure 2.29. One of the early RPC pilot plants. RPC photo, 1969.

By the spring of 1964, RPC's staff had risen to nine [38]. In the following year it had doubled to 18 [43], and by the next year it had almost doubled again, to 34 [55]. By 1968, the staff level was 51 [50].

With people, equipment, and facilities in place, Dr. Bursill noted in his Report to the RPC Council for June 1970 [73], that:

"the pipeline has been filling with ideas, experiments and projects to the point at which, now, we have more than we can deploy to industry."

Of course, not everything went according to plan and sometimes a company's problems were beyond RPC's ability to help. For example, at one time in 1969 Dr. Bursill was contacted by the Provincial Government's Department of Economic Growth to provide advice on *"some astonishing failures of a certain company and join an investigating committee…"* [74]. It was, however, too late and Dr. Bursill noted that *"Many attempts by our own staff to make useful contact with this company early enough to be helpful had failed and at no time did we have any explicit support of government. The whole point of RPC is to get at companies in time to be useful, not when the difficulties get dramatic."* Fortunately, there had been an abundance of cases in which RPC was called in early enough to be of material assistance.

In late 1969, RPC was well placed among Canada's six provincial research councils[16] (Table 2.1, see also Table 3.1 below).

Table 2.1. Canada's Provincial Research Councils in 1969 [75].

	Staff	Land and Buildings	Technical Equipment (at cost)	Annual R&D Work	R&D Spending Rank in Province
BC Research Council	91	9 ac 147,000 ft^2	$716k	$1,318k	3rd
Research Council of Alberta	208	12 ac 112,000 ft^2	$1,800k	$2,900k	2nd
Saskatchewan Research Council	83	10 ac 47,000 ft^2	$680k	$1,200k	1st
Ontario Research Foundation	250	90 ac 180,000 ft^2	$2,586k	$3,500k	1st
NB Research and Productivity Council	60	10 ac 65,000 ft^2	$1,500k	$850k	1st
Nova Scotia Research Foundation	60	10 ac 44,000 ft^2	$425k	$800k	1st

Dr. Shemilt's final comments, noted in the 1968/69 Annual Report [58] included the following:

"The initial years of planning, program development and staffing have finally led to a position where objectives in line with our establishing Act can be realistically set, and efficiently and consciously reached. In providing technical services and assistance, in achieving a well-recognized scientific capability and in the winning of confidence and respect from our peers and from those we serve, RPC is an increasingly mature though young organization."

Meanwhile, Dr. Bursill's thoughts, in December 1969, were on the future [76]:

"There are now signs… that the government of New Brunswick is seriously intent upon secondary industrial growth… the planning for major industrialization represents for RPC what will be, if the plans are realized, a new and distinct phase of its life. For this is what RPC has been preparing, organizing, staffing and equipping for… RPC is almost perfectly designed in outline to give the necessary service…"

It was 1969, and RPC was on its way.

---

[16] This data was collected before the launch of Quebec's industrial research council, CRIQ.

Table 2.2. Illustration of some of the changes experienced by RPC during *The Building Years, 1962 – 1969*. See also Tables 3.2, 4.1, and 5.1.

	1963	1969
Total Revenue	$50 k	$815 k
Provincial Grant	$50 k	$600 k
Earned Operating	$0 k	$215 k
Fixed Assets (mostly buildings, lab equipment and fixtures)	None reported	$2,430 k
Employees	3	56
Clients Served	0	118
Reference	[32]	[58]

3 THE GROWING YEARS, 1970 - 1983

> *The Science Council of Canada has "accepted that the provincial research councils are in essence national bodies and constitute the chief reserve of technological capabilities outside industry itself."*
>
> Dr. C. Bursill, Executive Director,
> Report to RPC Council, June 12, 1970

The Science Council of Canada's assessment [27], referred to above by Dr. Bursill, bears repeating here:

"the Research Councils and Foundations in the provinces of Canada are unique. While several other countries have institutes equivalent to our National Research Council, none has quite the same broad range of industrially and regionally oriented institutes in its states, provinces or districts… the six older Research Councils and Foundations appear to have become centres of competence which cannot be ignored. They have growth potential. They are quite effective and flexible in operation, and are generally closer to industry than are the universities. They are in a position to assess regional problems and to form an important link between local industry and the centres of competence in both provincial and federal departments and agencies. They are familiar with the problems of small companies, the more technologically oriented of which have perhaps more innovation and growth potential than most other companies. The Councils are, indeed, a Canadian resource … these institutes have grown and developed to an extent that their competence can be used beyond the borders of their respective provinces and for the benefit of Canada as a whole."

With the advent of the 1970s, RPC advanced *"further into the kind of service of greatest immediate impact in industry such as development, design, engineering, experimental manufacture and management services of many kinds"* [77]. At the same time, RPC experienced increased demand for technical consulting spanning a broad range of areas. In order to remain focused on productivity, RPC aimed to balance management services with technological services, and continued to emphasize contract work, which Dr. Bursill believed was an important measure of RPC's effectiveness [77]. He further noted [78] that:

> *"the work we do is increasingly of a consulting nature, whether concerned with the technology, broadly, of production or the technology of management and control. What this amounts to, on the one hand, is providing advice and assistance, backed by calculating, pilot-scale and experimental facilities, to governments at all levels and to industry; and on the other hand, it signifies diminishing in-house R&D in anticipation of a supposed need. In developing along these lines, we are responding to the demand and realities of our situation: innovation is beyond the financial and technical (indeed, of risk-taking) capabilities of the companies that operate in the region. There is much room for improvement, inventiveness and ingenuity and some of this we can hope to provide."*

Organizationally, Dr. Shemilt stepped-down as Chairman of the Council in 1969 (but remained a Member of Council for eleven more years, until December 1980. In his place, Mr. Kenneth V. Cox[17], was appointed RPC's second Chairman on May 13, 1970 [79,80] (Figure 3.1).

Meanwhile, RPC continued to focus on "mission-oriented research and technology" but now, as its programs matured, RPC was able to learn more from its experiences in the marketplace. Using this market knowledge, they began to increase their capabilities in promising areas while reducing efforts in areas that had proved to be less in demand [61]. In 1968/69, RPC achieved 31.5% of its total revenue as earned operating income – which was mostly contract revenue.

Contract revenue continued to rise, and RPC hit a milestone in 1971 by achieving over a million dollars ($1,026,329) in total revenue [80]. By 1972/73, earned operating income had risen to 36% of total revenue [77], with the level of government grant holding remarkably steady at $600,000 per year.

---

[17] At the time, he was President of the New Brunswick Telephone Company Ltd.

Figure 3.1. Mr. Kenneth V. Cox, a founding Council member and RPC's second Chairman. RPC photo.

Both the continuation and the amount of government support had to be justified (and essentially 're-sold') each year, of course. This was no trivial matter, but was successfully accomplished for many years running. In May 1973, for example, RPC's leaders "attended the provincial legislature's public accounts committee and were congratulated on our *very valuable contribution to the economy of New Brunswick*' by both sides of the house" [81].

By 1973/74 the numbers of clients served per year had reached the 200 level – with approximately 90% of the clients being based in New Brunswick [82].

Some examples of RPC work deemed newsworthy by the media in 1970 included [83]:

"a special team, the only one in Canada, working on the specialized field of fluid power research… a team of computer engineers developing a low-cost system for numerically-controlled machine tools… [and a team] reproducing experimentally the kind of minerals found in the province and in the process has synthesized several compounds of interest to the semi-conductor industry."

Figure 3.2 shows a particularly appropriate newspaper headline.

Figure 3.2. A treasured RPC news headline and feature article in the *Fredericton Daily Gleaner*, Sept. 12, 1970 [84].

At about the same time, a series of six articles on RPC was published in the *Fredericton Daily Gleaner* [71,84-88]. The first of these articles [84], quotes Dr. Bursill explaining what RPC does as follows:

"We are a problem-solving group… [with] one basic objective – to identify a problem and then solve it. We are not interested in academic research. We don't look for the reason why, although sometimes that can't be helped. We just want to know if it is so. We use a practical, pragmatic, down-to-earth approach here."

Illustrating the increasing variety of problems industry brought to RPC, the food product development work, and associated tasting panel, were both expanded to include pet foods in the early 1970s [80] (Figure 3.3).

Figure 3.3. A "distinguished member of RPC's consumer panel states preference for RPC-formulated pet foods." RPC photo, 1971 [80].

Figure 3.4. RPC's Dr. John Sutherland writing a computer program, on punch-cards, for a mainframe computer *circa* 1970. RPC photo [89].

As the 1970s advanced, RPC's programs were increasingly grouped by technical specialty into two groups of departments - one for "experimental" departments and one for "non-experimental" departments - and the Technical Information Service (TIS) was merged with Management Engineering [61,80,90]:

- ○ ***Chemistry & Food Science,***
- ○ ***Electronic Engineering, Instrumentation,***
- ○ ***Management Engineering,***
- ○ ***Mechanical Engineering,***
- ○ ***Minerals & Materials.***

Figure 3.5 (a and b) illustrate the proportions of different kinds of work (on an expenditure basis) performed in 1972/73 [77]. Figures 3.6 and 3.7 illustrate the work that was done in some of these areas.

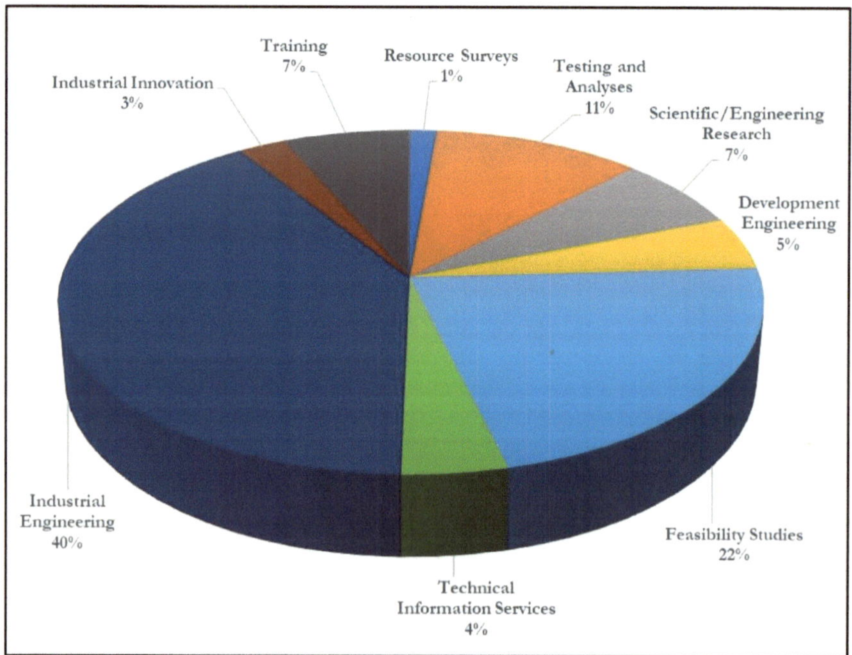

Figure 3.5(a). Illustration of the relative proportions of RPC work in 1972/73. This figure shows the data, displayed by type of work, on an expenditure basis. From data in reference [77].

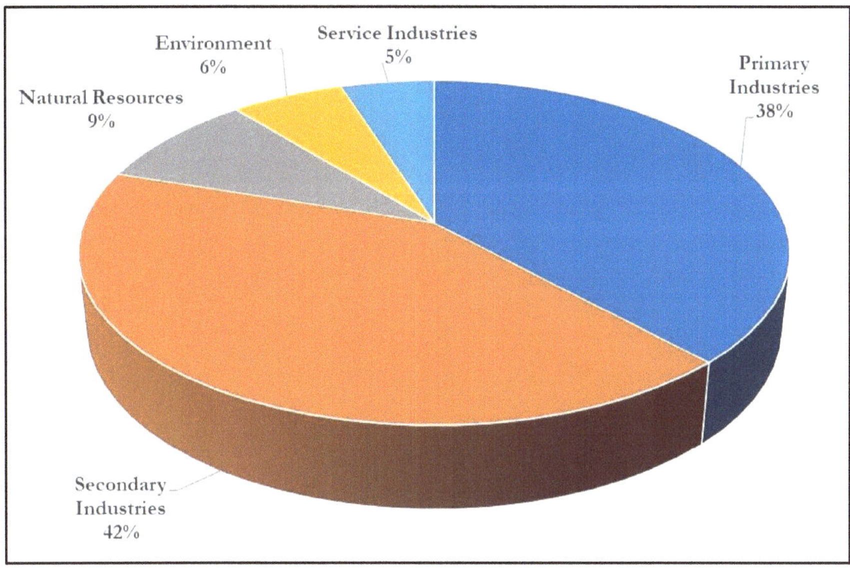

Figure 3.5(b). Illustration of the relative proportions of RPC work in 1972/73. This figure shows the data, displayed by type of sector served, on an expenditure basis. From data in reference [77].

Figure 3.6. Marjorie Gould conducting microbiological tests in a food processing plant (upper), a machine designed for H.B. Industries, and pilot plant recovery of by-products from industrial effluent (lower). RPC, 1972 [90].

Figure 3.7. From top to bottom: field, laboratory, and pilot plant work on the treatment of mine tailings. RPC's Gerry Ansell appears in the middle photo. RPC photos, 1972 [90].

Bricklin. Many had hoped that the futuristic-looking Bricklin SV-1 sports car would be a manufacturing success story in New Brunswick. The car's striking exterior, with its gull-wing doors (Figure 3.8), its unique plexiglass-reinforced-with-fibreglass composite car body (with colour-impregnated resin), and its leading-edge safety features generated much excitement when production began in 1974 [91]. The car bodies were made in a plant in Minto, N.B., with the vehicle assembly being done in a plant in Saint John [91]. RPC had provided Bricklin with engineering support related to improving the plexiglass-fibreglass car bodies [92]. RPC's work in support of Bricklin gained public attention when Bricklin was having financial difficulties in 1975, and it came out that they owed RPC money [93]. Although some of the headlines, like "Bricklin Owes Money to RPC," were eye catching the amount actually owed to RPC was quite modest at a reported $29,000 [93], and the account with RPC was settled within a few more months [94]. The provincial government was not as fortunate, however, having invested and lost an estimated $23 million by the time production of the cars ceased in September 1975 [95,96]. Of the estimated nearly 2,900 cars manufactured [95], numerous examples are still in operation.

Figure 3.8. A Bricklin SV-1. Photo courtesy of C. Ziemnowicz, 2012, Wikimedia Commons [97].

RPC's pilot plant and experimental manufacturing facilities were increased with the construction of its next major facility, the Production Transfer Pilot Plant (PTPP). The new RPC plant was announced in the March 1974 Throne Speech, in the Provincial Legislature, saying that to encourage businesses to exploit ideas the pilot plant would be built *"to test the economic and technical viability of new products with "successful product lines transferred to the private sector for commercial operations"* [98]. This would be building #5, for RPC and it was opened in August 1975 [99,100]. It comprised 20,000 sq. ft. and high-head height of 20 ft. (Figure 3.9).

Part of the strategy for the PTPP was to use it, not just to develop and demonstrate technologies, but to develop the 'craft skills' that would also have to be transferred to the private sector in order for the commercialization of the new technologies to be successful, especially in the cases of small- and medium-sized businesses. As Dr. Bursill explained it [99]:

"The whole problem turns upon the notion that technology, unlike science and scholarship, is transferred by agents not agencies. Each craft operation has roots in practices which have been built up, sometimes over very long periods, which are absolutely specific and which may be unknown to management — a series of tricks learned by trial and error; if they are not passed on, they must be learned again, either rationally by research, which may be enormously expensive and time-consuming, or by making the errors in conditions of start-up when they are no less costly. So it seems that… there might be a transfer cost that is inherently unbearable by industry…"

With processes developed and demonstrated at PTPP, RPC's approach would be to develop the necessary craft knowledge and then *"it must be a gift to the company"* which, for the public, *"would be priceless and which constitutes a wholly new or modified development base, equivalent to grafting development on to natural resources"* [99].

Early experiences with this approach had been successful [99]:

"We find that we are much turned to for some capabilities missing elsewhere. It may happen that a particular shape must be cut from metal, for example, and it is discovered that only RPC can combine the necessary engineering with craft to produce the required order of accuracy… But we do discover that in such cases we act as pilots… for when we establish the method, it suddenly becomes easy and can be turned back to 'usual channels.'"

In other words, with sufficient training and demonstration of the craft, the clients could then take up the processes on their own.

Figure 3.9. RPC Building #5 (left of centre), housing the PTPP pilot plant and experimental manufacturing facilities opened in 1975. RPC photo.

This kind of attitude made for a large number of highly satisfied, and returning, clients. So much so, that some companies began to factor proximity to RPC in their planning to build or relocate facilities. For example, in 1975 Kasper and Richter (K&R) Precision Instruments, a West German manufacturer set-up a plant in the Fredericton Industrial Park to manufacture pedometers, map measures, and related products. They explained that they chose Fredericton for their manufacturing plant site because, in part, *"the Research and Productivity Council is located here [which] is a remarkable advantage, since we appreciate the valuable technical aid the council can provide"* [101]. This kind of client behaviour provides a strong indicator of the regional economic impact of an RTO's work.

To further enable and assist the development of process plus craft plus product that could then be transferred to private companies, RPC created its first subsidiary company, Innovent Ltd. (see Appendix 8.4). Innovent would be the agent for stimulating PTPP activities and seeing them through, if successful, to commercialization. From this was born Galleon Ware.

Galleon Ware. The first large-scale venture under the new PTPP strategy involved the development and design of ceramic tableware for commercial pottery, based on the use of local clays. Although this activity was begun under the Innovent subsidiary in 1974, it soon showed sufficient promise that its activities were moved to a new, dedicated RPC subsidiary, Galleon Ware Ltd., which was incorporated in April 1975 (see Appendix 8.4). Galleon Ware would not only do the applied research, development, and demonstration, but also full-scale production, marketing and sales.

Galleon Ware commenced commercial production in 1975/76 with its first product line placed on the market for trials across Canada, a second product line being sold to a client in the U.S., and two additional product lines under development [92,102] (see Figures 3.10 and 3.11). Galleon Ware Ltd. held the trademark and the pottery identification mark for products produced under its own name (Figure 3.11).

As 1976 progressed, more product lines were developed, including a line of oven-to-tableware line [103]. Galleon Ware was formally launched, under its own brand, in the US at the Atlantic City China and Glass Show in Atlantic City January 9-13, 1977 [104,105]. In Canada, Galleon Ware was being sold from local stores, like Fredericton's J.D. Creaghan Ltd., to the nation-wide chain Henry Birks and Sons [106]. Richard P. Creaghan, manager of the Fredericton Creaghan's store was quoted as saying that Galleon Ware produced *"the nicest coffee mug I've ever held,"* and that *"the store sold [Galleon Ware] coffee mugs every day and found that the coffee sets were in demand as gift items and for winter entertaining"* [106]. Meanwhile, RPC even had its own 'factory outlet' at the pilot plant building (Figure 3.12) [107]. An example of a Galleon Ware product line is shown in Figure 8.4. (in Appendix 8.4).

In the spring of 1977, however, disaster struck. Galleon Ware's main distributer reported *"overstock situations existing throughout the dinnerware market with stores offering discounts as high as 40%, even on prestige products by Wedgewood, Doulton, Derby and Mikasa"* [108]. Not only were large-scale china producers entering the international market, but low-cost alternative products like Corelle Ware[18] were gaining market share. Unable to compete with the new, lower market price regime, Galleon Ware turned to the giftware market, where it was hoped that the preponderance of custom-made products would be less susceptible to market fluctuations [108]. Unfortunately, this strategy was only moderately successful and the business had become marginal at best for RPC.

---

[18] Corning Glass Works' product Corelle Ware, for example, was an inexpensive, light-weight competitor that was oven, dishwasher, and microwave safe. It is made of tempered, laminated glasses and is still being produced by Corelle Brands.

Figure 3.10. Making Galleon Ware pottery in RPC's pilot plant facility. RPC photos, 1975-1976 [92].

Figure 3.11. Examples of Galleon Ware Pottery made in RPC's pilot plant facility. The pottery identification mark for Galleon Ware is shown in the lower right. RPC photos, 1975-1976.

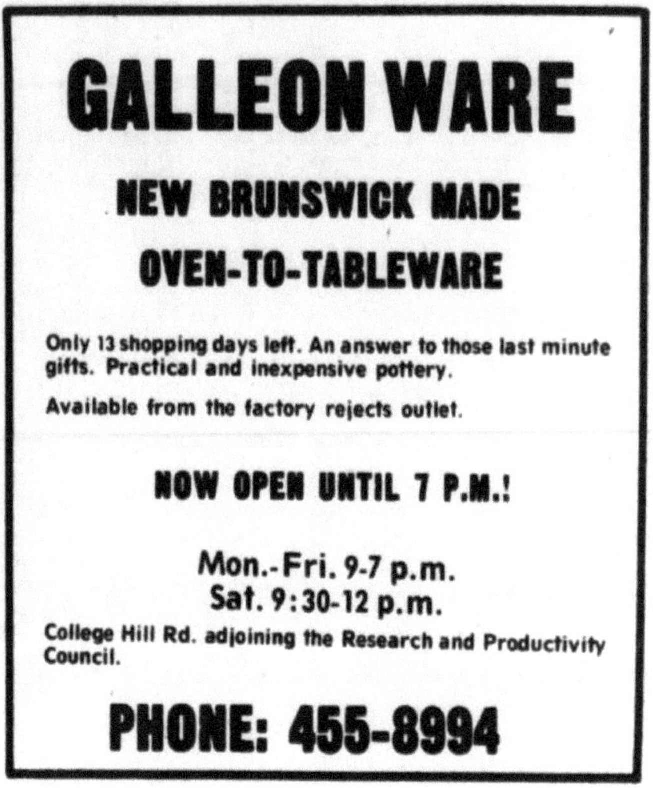

Figure 3.12. A 1976 advertisement for Galleon Ware from RPC's 'factory outlet' at the pilot plant facility [107].

Galleon Ware Ltd.'s own production was ceased in September 1977 [109]. However, RPC's goal of creating, demonstrating, and ultimately selling the business to the private sector was accomplished, as it was sold to Accord Ceramic Industries Ltd. [110] in October, 1977. Accord's plant went into full production in the spring of 1978 [110].

In the end, the PTPP strategy was well demonstrated via the Galleon Ware initiative. Some pottery lines had been successfully transferred to the private sector and, ultimately, the core business had been transferred to a local New Brunswick company.

As a postscript to this initiative, an internet search shows that the originally produced Galleon Ware pottery lives on to the present day, and sets and individual pieces can still be found for sale on such sales sites as eBay, Kijiji, and Etsy [111].

Energy Efficiency. The energy 'crisis' of the 1970s, fuelled by the sharp increase in energy and fuel prices, led to much concern and focus on energy consumption across Canada, and New Brunswick – and therefore RPC - was no exception. Hugh Drummond, head of RPC's Engineering Department, and others from the council gave many presentations to Chambers of Commerce, and other public forums, on energy conservation and alternative energy technologies (see, for example, references [112,113]). In this area, RPC employees tried to communicate more broadly than usual and get reliable technological information to the public at large, as well as their more usual audiences of small-, medium-, and large-scale industry.

By 1979, several provinces launched mobile energy audit programs, (MEAP) popularly referred to as 'energy-bus' programs, to help companies reduce their energy consumption, and RPC was asked to implement one for New Brunswick. With funding from the federal and provincial governments, RPC was able to offer a free service to qualifying companies. RPC's energy bus (Figure 3.13) would visit the companies, conduct an energy-use audit and analysis, and then identify and recommend energy efficiency opportunities for the company along with an estimate of the potential energy savings. By October 1979, 107 New Brunswick companies had participated, over $4 million in energy bills were audited, and over $800,000 in potential energy savings identified [114]. RPC's energy audit programs continued through the 1980s, and past the termination of the federal funding programs [115].

In 1981 RPC was contracted to deliver a federal-provincially funded Small-Scale Conservation and Renewable Energy Demonstration Program [116]. This program was made broadly available (Figure 3.14), including to individuals, not-for-profits and institutions, small businesses, and municipalities. Among the kinds of eligible projects were [116]:

o *Energy conservation*, including retrofits, waste heat recovery, heat storage, and car pooling,
o *Biomass energy*, including wood and wood-chip space- and water heating, and alternative fuels,
o *Solar energy*, including solar domestic hot water heaters, passive solar home construction and retrofits, and wood/crop drying,
o *Wind energy*, including remote electrical generation using packaged units, and
o *Micro-hydro* electrical energy generation.

Figure 3.13. A 1979 advertisement for RPC's 'Energy-Bus' Program [117].

An energy conservation example is the 1983 creation of a federal-provincially funded Combustion Test Demonstration Facility at RPC, which would be used to evaluate the energy efficiency of residential combustion heating equipment (and to train furnace service personnel) [118]. It was hoped that these services would lead to heating system improvements, retrofits, and replacements creating a 15% average increase in energy efficiency [118].

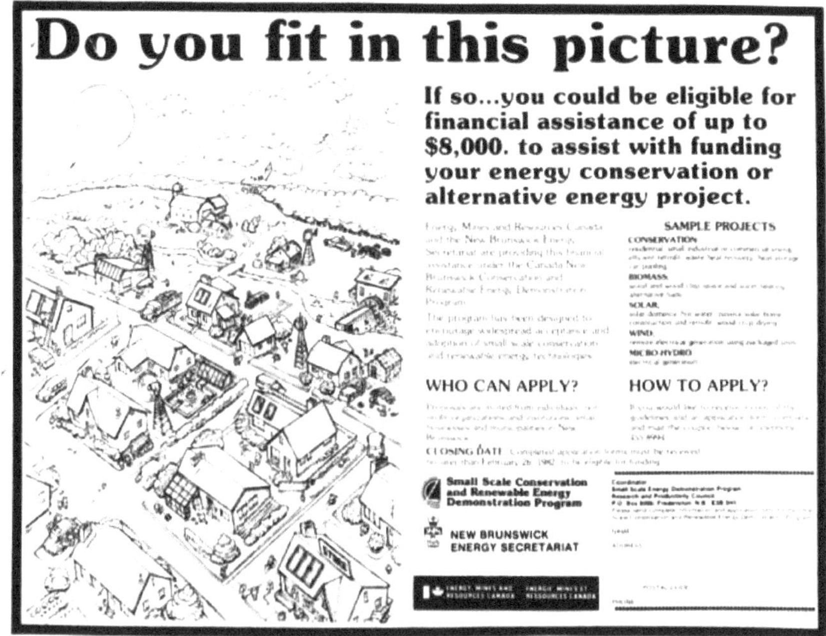

Figure 3.14. A 1981 advertisement for RPC's Small-Scale Conservation and Renewable Energy Demonstration Program [116].

Another form of alternative (but not renewable) energy, besides those listed above, is nuclear power. Encouraged by N.B. Electric Power Commission and Atomic Energy of Canada Ltd. (AECL), RPC began its nuclear involvement in 1976 with work in non-destructive evaluation and metallurgy, and ultrasonic flow metering, for the planned Point Lepreau nuclear power station [119].

RPC's first contract with AECL was in early 1976, for $100,000. Further contracts ensued, with the quality assurance inspection work being *"handled jointly between officials of the provincial power commission, of Atomic Energy of Canada Ltd. and the New Brunswick Research and Productivity Council"* [120].

It wasn't long before RPC's work had a positive impact on the plant. In one example, an RPC team designed the machine to detect vibrations in heavy-water-carrying boiler tubes that led to their replacement in the plant [121]. This was not a trivial issue: in the end, all four steam generators had to be retubed before the billion-dollar-plus project could be safely completed [122]. The knowledge and spin-off technology gained by RPC on the Point Lepreau projects were used to serve the nuclear industry at large including, for example, RPC work for the Gentilly 2 station in Quebec, Pickering B in Ontario, and Wolsung 1 in South Korea [119].

RPC's TIS program services continued through the 1970s and '80s (Figure 3.15), but in 1981 they were consolidated under the new NRC Industrial Research Assistance Program (IRAP), which NRC contracted RPC to deliver throughout New Brunswick[19] [123].

A 1979 media feature on RPC [2] extolled some examples of its work and impacts of the times (Figure 3.16) and enabled Dr. Bursill to say a few words about RPC's approach:

"The aim of RPC is to help companies use what they've got to answer problems they face. It is easy for a company to solve a production problem 'radically' but few companies, even the larger ones, have the time or are willing to take the risk. Most companies the size we have in New Brunswick can't afford a major turnover. We have got to find for them a way of doing their job quickly and efficiently in the context of the machinery they already have."

This approach clearly resonated with RPC's industrial clients. Not only did the numbers of clients grow, but many became long-term clients. For example, RPC's first publicly recorded industrial client list (1965/66 [55]) comprised the following:

- Brunswick Mining & Smelting Co.
- Campbell Soup Co.,
- General Foods Corp.,
- McCain Foods Ltd.,
- Plasticraft Ltd.,
- Saint John Shipbuilding & Dry Dock Co.,
- St. Stephen Textiles Ltd.,
- T.S. Simms & Co.,
- Sugar Research Foundation.

Some twelve years later, seven of these nine were still RPC clients, as recorded in its last public listings of clients (1976/77 and 1977/78) [124,125][20].

---

[19] Until 2000 through 2002, when NRC transitioned to delivering this service directly, with its own staff.

[20] McCain Foods wasn't actually listed in the two reports referenced here, but were still an RPC client by 1978 and, indeed, have remained so to the present day.

Figure 3.15. Forest products companies being assisted by RPC's Technical Information and Industrial Engineering employees including Bob Reid and David Kersey (upper) and Don Young (lower). RPC photos, 1976-1977 [124].

18 The Daily Gleaner, Friday, June 22, 1979

Over 200 Companies Used Expertise In 1978

Research, Productivity Council Serves N.B.

"We get together with the client and sit around and talk until we have found out what he has in mind. We draw up a program and present him with a proposal. If he accepts it, we are now on our way to solving the problem."

Getting the 'bugs' out of a system to make it efficient and cheap is one of the prime targets of any competitive business.

And more than 200 of New Brunswick's approximately 11,000 companies used the Research and Productivity Council last year for that purpose.

A limited survey of the companies, whose names appear as clients in the RPC's annual report, resulted in an almost unanimous loyalty to the council, its staff and its purpose in New Brunswick's economy.

Some of the companies on last year's list have since gone bankrupt, such as Marven's Ltd. of Moncton, a confectionary manufacturer, or Thermofoam Mfg. Ltd. of Fredericton which has since been bought out by Truefoam Ltd. of Capital Co-Op.

**Stories By
Gail Dugas
Staff Writer**

But, according to Commerce and Development assistant deputy minister Pat Blanchard — who also sits on the board of the council — the council cannot act as a cure-all to an ailing company and can only deal with specific problems.

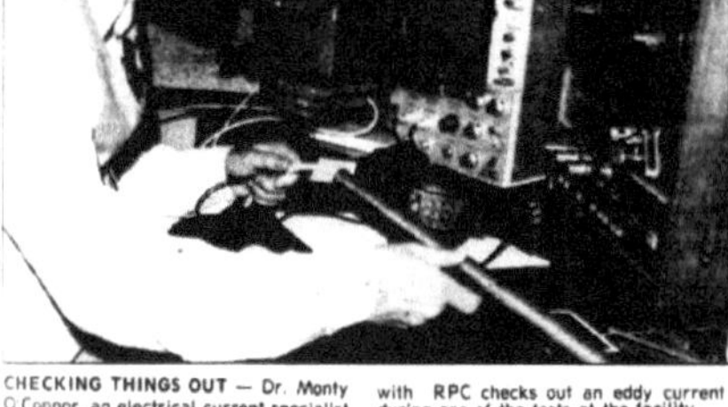

CHECKING THINGS OUT — Dr. Monty O'Connor, an electrical current specialist with RPC checks out an eddy current during one of the tests at the facility.

PROBING ELECTRONS — Some testing and research at the RPC involves the use of machinery such as the electron microprobe used by Dr. J. K. Sutherland.

RPC Uses Scientific Technology To Aid Industry

The necessity of employing science and the latest in technology is almost a fact of life in business these days — no matter how small the enterprise.

Yet as in the rest of Canada, less than three per cent of New Brunswick industries have the capacity or facilities for research and development.

Almost 16 years ago the provincial government set up the New Brunswick Research and Productivity Council to help manufacturing, processing and other New Brunswick industries deal with competition. As RPC director Dr. Claude Bursill puts it: "The aim of RPC is to help companies use what they've got to answer problems they face."

It is easy for a company to solve a production problem "radically." Dr. Bursill says that few companies, even the larger ones, have the time or are willing to take the risk.

Most companies the size we have in New Brunswick can't afford a major turnover. We have got to find for them a way of doing their job quickly and efficiently in the context of the machinery they already have." Dr. Bursill says he likes to call that process "technical elegance."

As quality control regulations get more stringent and competition gets fiercer, trying to run a business "the old way" can be needlessly expensive if not catastrophic.

The RPC office on Kings College Road has 75 people on staff — 40 of them professional engineers or scientists working in the centre's five departments. The departments are set up to handle frequent problems encountered by segments of the province's business sector.

Dr. Gordon Brown, a food scientist trained in Britain, heads up the chemistry and food science division; Dr. Andy Mitchell is director of the engineering department which handles diverse problems such as striking special tools to grinding finely-tuned machine parts; a special department to specifically deal with management, productivity and market problems is headed by Dr. Peter Lewell while Dr. Donovan Abbott is director of the largest department — minerals and materials.

The council handles approximately 2,000 jobs a year. Most of the companies they deal with employ an average of 11 people.

Dr. Bursill describes most of the work that needs to be done for New Brunswick companies as "kitchen sink chemistry — your basic things." It's not the same as finding a new galaxy but it is work that has to be done.

Not all the work is mundane though. Dr. Bursill says as the council's reputation grows in certain areas, research contracts come in from around the world. Dr. Brown recently spent time in the Sudan researching possible uses for a certain type of tree.

Construction of the Point Lepreau nuclear power plant has been a boon to engineering research in this area, according to Dr. Mitchell whose department has been working on a contract basis with the New Brunswick Power Commission and Atomic Energy of Canada Ltd., to make some adjustment to the plant design. It was an RPC team that designed the machine to detect vibrations in boiler tubes which are being replaced in the plant.

(See SCIENCE Page 19)

Figure 3.16. Clips from a 1979 feature article on RPC [2,121].

In 1981, RPC was still well placed among Canada's eight provincial research organizations, according to a Science Council of Canada survey [126] (Table 3.1, see also Table 2.1 above).

Table 3.1. Canada's Provincial Research Organizations in 1981 [126].

	Staff	Provincial Government Grant	Contract Revenue	Annual R&D Work
B.C. Research	163	$1,375k	$5,751k	$7,528k
Alberta Research Council	506	$15,000k	$14,894k	$26,376k
Saskatchewan Research Council	244	$3,065k	$7,106k	$9,806k
Manitoba Research Council	40	$2,823k	$558k	$2,292k
Ontario Research Foundation	438	$3,428k	$12,563k	$17,106k
Centre de Recherche Industrielle du Québec	314	$7,500k	$4,486k	$13,466k
N.B. Research and Productivity Council	90	$600k	$3,778k	$4,034k
Nova Scotia Research Foundation	128	$1,300k	$2,651k	$3,795k

Mineral Resource Development. As RPC ramped up its activities in mineral resource identification and development, a dedicated subsidiary company for this purpose was created: Minuvar Ltd[21].

An early Minuvar success was in limestone development. There are a number of deposits of limestone[22] in the Petit Rocher area, which is about 30 km north of Bathurst, New Brunswick. Some quarrying had been carried out as far back as the 1880s, then again in the 1930s through early 1960s (including by Elmtree Resources Ltd. and Elmtree Limestone Cooperative Ltd.). However, due to the non-uniform distribution of the limestone in the region each quarrying operation only accessed a tiny fraction of what is now thought to be a limestone resource capable of supporting quarry operations with reserves in the order of several million tonnes [127]. Despite these efforts, and estimates, subsequent private sector drilling by other companies was unsuccessful.

In order to revive commercial interest in the limestone opportunity, the New Brunswick Department of Natural Resources and Energy, and RPC

---

[21] Minuvar Ltd. was incorporated in New Brunswick in April 1975. It is still held by RPC to the present day (see Appendix 8.4).

[22] In the La Vieille and LaPlante formations.

(through its Minuvar subsidiary) joined forces to bring modern geophysical techniques – in this case, terrain conductivity and Gamma-ray radiometry - to the problem of identifying the most promising limestone masses in the region.

In 1974, 16 mining claims were staked, mapped and sampled by Minuvar Ltd. in the Elmtree area, and in 1975, additional mining claims were added in the LaPlante area. Even with the assistance of modern geophysics, a subsequent drilling campaign still only intersected good quality limestone in two out of eighteen drill holes, but it was enough. Elmtree Resources Ltd. acquired the Elmtree property in 1978, carried out confirmatory trenching and mapping, established a 246 ha mining licence and a large crushing, screening, and storage facility. Most of the limestone produced originally went to Brunswick Mining and Smelting for their lead smelter at Belledune, but over time other limestone products were developed and sold to others, including a flue-gas-desulphization product for two regional power plants at Belledune and Dalhousie. This operation ran successfully until the early 1990s, at which time Elmtree Resources Ltd. shifted to another of the high-quality limestone resources that had been identified, this one some 27 km to the southwest, near Sormany [127].

Another area undertaken by Minuvar was in manganese development. There is an iron-manganese deposit, known as the Woodstock Property, located approximately 6 km west of the Town of Woodstock, N.B. The ore was originally mined in the late 1800s for its iron content, which was sold to local smelters [128]. The manganese potential became better appreciated through various Geological Survey of Canada and private sector assessments conducted in the 1930s through '60s, some of which estimated the manganese resource in the Woodstock deposits to be of the order of 200 million tonnes. In the early 1970s, Mandate Refining Co. unsuccessfully attempted to develop the deposits, and eventually abandoned its claims [128].

Here again, in order to revive commercial interest in the manganese opportunity, RPC (through its Minuvar subsidiary) took up the claims, and between 1976 and 1986, conducted geological mapping, geochemical sampling, and geophysical surveys (magnetometer and very-low-frequency electromagnetic (VLF-EM) surveys) of the area [128-130].

This was enough to rekindle interest in the deposits, and additional mapping, trenching, and surveying was conducted by private-sector and government organizations through the late 1980s and 1990s. Along the way the mineral claims were bought and sold several times, ultimately vesting with Canadian Manganese Company Inc. (CMC). At the time of writing, a pre-feasibility study for a manganese mine was at the planning stage.

Mineral Processing. RPC's mineral processing activities rose to new heights in the late 1970s and early 1980s (Figure 3.17), particularly with its sulphation-roast-leach (SRL) electrowinning technology for the processing of low-grade copper-lead-zinc ores. With this process, it was hoped that RPC could make it possible to increase the profitability of existing New Brunswick mines, and also to enable the profitable mining of marginal grade ores [125].

"The RPC process overcomes the metal extraction difficulties that plague New Brunswick ores and commercialization of this technology was considered a key to any substantial expansion of the New Brunswick mining industry" [131].

By 1983, RPC would be able to boast [132]: "30 patents in 12 countries of the world" for this made-in-house technology (some of which are listed in Appendix 8.8).

Sherritt Gordon Mines Ltd. was an early supporter of this work [133]. In 1979, a consortium was formed, with four companies, to conduct a preliminary planning and engineering design study for a mini-pilot plant [134,135]. This was front-page news in the region [135] and led to construction and commissioning of the plant, Figure 3.18, which was used to develop, test, and demonstrate the process for ores from different companies. In 1981, for example, such mini-pilot plant testing was conducted for Anaconda Minerals Co. to demonstrate the applicability of the SRL process to their 'Caribou Ore' in northeastern New Brunswick [123].

The mini-pilot plant demonstrations were sufficiently successful to warrant the construction of a larger pilot plant. Federal-provincial funding for the plant was announced in August 1983 [136], and for this RPC created another new subsidiary company. Enhanced Recovery Systems (ERS), with its head office in Chatham, N.B., was incorporated[23] September 7, 1983. Its purpose was to:

"acquire, construct, and operate all phases of a 10 ton per day pilot plant to test the RPC Sulfation-Roast-Leach-Electrowinning (RPC S-R-L) Process and the RPC Brine Leach Process for improving the extraction of metals and other saleable products from complex base metal ores typical of New Brunswick" [137].

For the pilot plant site, ERS purchased a 17-acre, former Ciba Geigy plant-site at Morrison Cove (near Chatham, N.B.) and renovated it for this purpose (Figure 3.19). Once the original metal extraction processes were successfully developed and demonstrated, it was hoped that other custom or contract test opportunities of suitable scale would arise. The new facility was expected to be ready by end of 1984 (see Chapter 4).

---

[23] ERS is currently inactive. See Appendix 8.4.

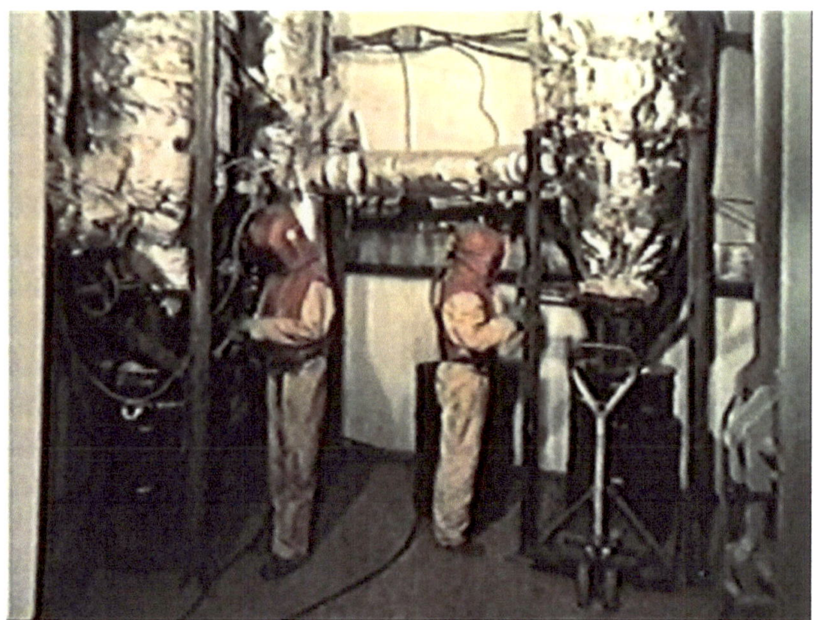

Figure 3.17. Mineral-processing pilot plant operation *circa* 1978. RPC photos, [89].

Figure 3.18. Mini-pilot plant project demonstration of RPC's sulphation-roast-leach (SRL) electrowinning technology for the processing of low-grade copper-lead-zinc ores. (Left to right: RPC's Dr. Roy Boorman, Ross Gilders, and two guests. RPC photo, 1978.)

Figure 3.19. RPC subsidiary ERS' pilot plant facility at Morrison Cove (near Chatham, NB). The first projects involved a 10 ton-per-day pilot plant development and demonstration of RPC's Sulfation-Roast-Leach-Electrowinning and Brine Leach processes. The ERS subsidiary's logo is shown below the photo. RPC images 1984 [137].

Another major RPC project in mineral processing was also an energy project. RPC had taken on projects related to coal utilization since about 1978 [138], and extended this into working on oil shales. The resource potential was, and is still, huge: the amount of synthetic crude oil potentially recoverable from New Brunswick's Albert County oil shales has been estimated to be billions of barrels [139,140].

RPC's coal and oil shale development work quickly came together in the form of fluidized-bed combustion of mixtures of oil shale with coal. The basic concept was to use the oil shale to enhance steam production from conventional coal-fired boilers, to generate electricity. Among the proposed benefits of this process were the reduction of sulphur emissions (from New Brunswick's high-sulphur coal) and the generation instead of by-product gypsum, which could also be marketed [141].

This work led to the construction of a $35 million Circulating Fluidized Bed (CFB) demonstration plant at Chatham, N.B. by the N.B. Electric Power Commission (~1982-1986) [142].

Figure 3.20. Mini-pilot-plant process development for retorting Albert County oil shale. RPC photo, circa 1980.

The foregoing examples of RPC programs of the 1970s and early 1980s represent only a few of many initiatives. A 1983 RPC newspaper advertisement listed [143] the following areas of company focus:

o **Energy** - Coal and oil shale utilization, combustion technology, biomass fuels, domestic furnaces, pyrolysis, battery storage, conservation, fuel use management.

o **Engineering** - Non-destructive testing, field inspection services, inspection procedures, metallurgy, materials selection, welding technology, corrosion investigations, instrumentation.

o **Chemistry** - Analyses, process control, occupational health and safety, alternate pest control strategies and communicational chemistry, quality assurance, referee determinations.

o **Ecology** - Pollution problems, pesticide residues, bioassays, land rehabilitation, fish resource assessment, catch preservation, aquaculture, socio-economic impact studies.

o **Food Science** - Resource utilization, by-product recovery, biotechnology, quality assurance, industrial processes, product formulation, consumer testing, microbiology.

o **Mechanical Technology** - Design and prototype manufacture, fabrication of electrical-electronic control systems, precision machining, computer aided design and machining.

o **Mineral Processing Development** – Pyro-hydrometallurgical processes, fluidized bed technology, innovative processing, beneficiation, computer aided process analysis and simulation.

o **Productivity Improvement** - Manufacturing opportunity evaluation, sector development planning, technical information services, industrial engineering, technology demonstration and training.

An example of RPC's work with N.B.'s cultivated mussel industry is the pilot-scale longline cultivation trials held with local fishermen during the open-water season of 1981/82, which led to an instruction manual that was developed for the N.B. Department of Fisheries. The manual, titled **Handbook for Mussel Cultivation**, describes rearing methods, equipment requirements, and the demands in terms of human resources and finances for a successful commercial venture [123].

On a different note, RPC received its own Coat of Arms and Badge (Figures 3.21 and 3.22). They were granted by Letters Patent of the Kings of Arms[24] on September 20, 1978 [144] and are registered with the College of

---

[24] Grants of Arms are made by the Kings of Arms, using powers delegated to them by the Crown. The Earl Marshal, the Duke of Norfolk issues a Warrant to permit them to do so.

Arms, London, England[25]. The meaning of the Coat of Arms design is as follows:

- o The shield of black (*sable*) and white (*argent*) bars symbolizes the final output from RPC, which is nearly always words in print - black on white - the conversion of ideas, laboratory work, calculation etc. into useful information or know-how and do-how that must be transferred to a client mainly through reports. The 'barry' counter-change of black to white and white to black on each side of the shield and a further counter-change in the outer ring (roundel) symbolise finding a way through a maze. That is to say: solving problems which is so much to do with the life of RPC.
- o In the centre of the shield is a roundel of red (*gules*) symbolizing the Council in session at a round council table and the leopard's head in gold (*or*) within it represents the authority of Council as a body at the centre of our activity, hinting, too, at the connection to the Crown.
- o The crest of a *lymphad* echoes the New Brunswick coat of arms and the gold lion amidships again hints at a Crown connection. The red flags are for graphical balance and give energy to the ship.
- o The two supporters are animals common in our woods, the owl and the raccoon symbolizing wisdom and ingenuity respectively. They are in more or less natural colours, that is, in heraldic language: "proper."
- o The motto *"exitus acta probat,"* meaning[26] "the end proves the deed" is intended to convey that the results prove our work, i.e., it is our results that matter to our clients.

The meaning of the Heraldic Badge design is as follows:

- o The two fiddleheads represent freshness and growth, as well as some essence of the Province,
- o The fiddleheads interlace a second-order Möbius strip[27], which was described as a single quartic surface with equation $y^4 + x^2 (y^2 - z^2) - 2axyz - a^2y^2 = 0$. This symbolizes the inventive ingenuity of the Council.

---

[25] Prior to 1988, Canadian coats of arms had to be granted by the Crown in England. Since 1988, they have been granted by the Crown in Canada.

[26] The Latin phrase *exitus acta probat* is from the Roman poet Ovid. An alternative, fairly literal translation is: "the result validates the deeds."

[27] The Mobius strip in this badge was adapted by permission of the owners from an existing family arms.

Figure 3.21. RPC's Coat of Arms.

Figure 3.22. RPC's Heraldic Badge.

In September 1983, Dr. Bursill retired from RPC [39]. As RPC's leader during *The Building Years* (1962 – 1969) and *The Growing Years* (1970 – 1983), Dr. Bursill guided RPC through its evolution from a granting agency to an independent research establishment with its own dedicated scientific, engineering, and technological staff, complete with its own laboratories, pilot plants, and subsidiaries. By demonstrating an ability to quickly deliver practical results for industry problems, while maintaining a modest but dependable base of funding from the province of New Brunswick, RPC realized significant increases in 'earned operating revenues,' as he called them (Figure 3.23).

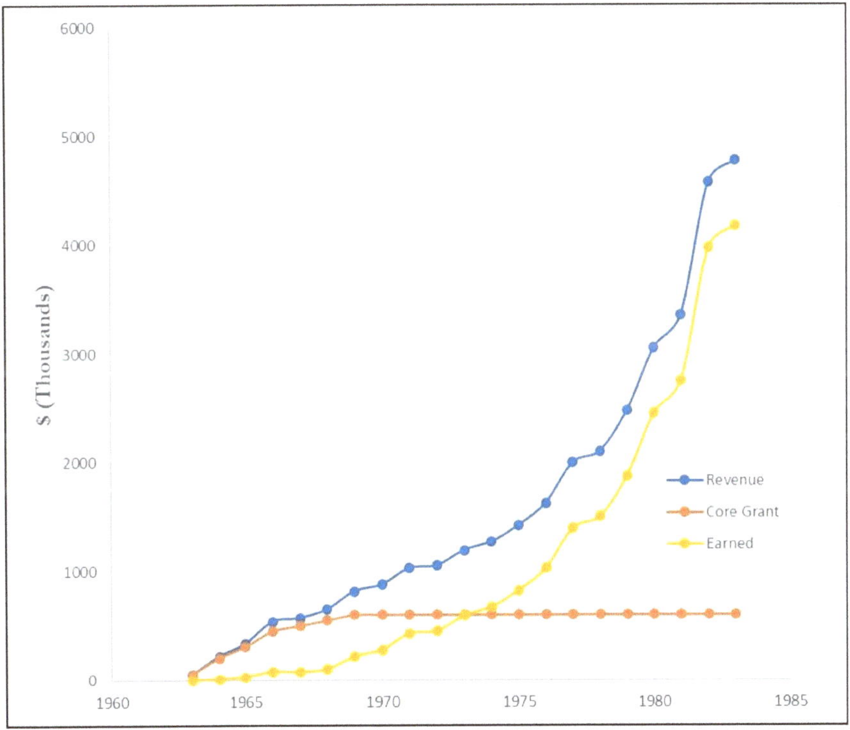

Figure 3.23. RPC revenue growth to 1983.

In a media interview, with characteristically pointed understatement, he summarized some of the challenges RPC had faced in its early years [39]:

"The conditions under which the council operates could have doomed it to failure. Our grant has been reduced, we don't compete with industry… the auditor general thinks we should not be allowed a reserve [fund] and we can't borrow [money]. It combines to totally hamstring one and yet despite this, we have been ingenious enough to live on… [and be] good for the province."

RPC did more than live on and be good for the province. In *The Building Years*, RPC evolved from an initial government investment of $50,000 to annual revenue of $815,000 (and a staff of 56) in only six years. In *The Growing Years*, RPC was able to realize average revenue growth of about 13% per year and the number of employees had nearly doubled to just over one hundred (see Figure 3.23 and Table 3.2).

It was 1983, and RPC had matured.

In a 1983 review of the provincial research organizations of the time Le Roy and Dufour concluded that [18]:

"… their common purpose is to make available the fruits of research likely to be beneficial to the provinces and to the nation as a whole, with particular emphasis on industrial development. This they do by undertaking research and development work and by making available technical information, advice, and know-how, as well as by providing analytical and testing services. In assuming this role, they serve as veritable research arms for thousands of small and medium-sized enterprises, an important segment of industry for employment and the creation of wealth in the country. They also serve as significant instruments in realizing both federal and provincial strategies for technological development by maintaining and enhancing technology transfer, industrial productivity and innovation."

These features applied to the RPC of the time as well as to any of its sister Provincial Research Organizations (PROs) in Canada.

Table 3.2. Illustration of some of the changes experienced by RPC during *The Growing Years, 1970 – 1983.* See also Tables 2.2, 4.1, and 5.1.

	1970	1983
Total Revenue	$877 k	$4,774 k
Provincial Grant	$600 k	$600 k
Earned Operating	$277 k	$4,174 k
Fixed Assets (mostly buildings, lab equipment and fixtures)	$2,805 k	$1,509 k
Employees	59	103
Clients Served	155	>200
Reference	[60]	[145]

4 THE COMMERCIAL YEARS, 1983 - 2004

In October of 1983 Dr. Roy S. Boorman[28] was promoted to become RPC's second Permanent Head [146]. He served in this capacity, as Executive Director of RPC, until 1992 (See Appendix 8.3).

Within two weeks of his appointment, the media was reporting: "***RPC Entering Most Critical Stage. Dr. Boorman Plans Different Direction***" [147]. In an interview on RPC's future direction he indicated that RPC needed to operate more like a business and continue to grow revenue from contracts. This was summarized by a reporter [147] as follows: in the previous era, RPC had "*devoted much of its time… to developing its expertise. Now, Dr. Boorman is out to sell this expertise.*"

At least part of the change in direction would have been due to the fact that Dr. Boorman assumed the leadership in the same year (fiscal 1983/84) as the provincial government reduced its annual provincial grant to RPC by half, from its longstanding level of $600,000 down to $300,000 [148].

It is easy to imagine that the organization's strategy would have been to (a) attempt to persuade the province to restore the previous funding level in future years (which was successfully accomplished, beginning with the 1984/85 fiscal year) and, (b) focus on building contract revenues in case that didn't work. Both strategies were successful. The provincial grant was restored to its former level in 1984/85 [149] and then increased to $800,000 in 1987/88 [150]. This was further bolstered in 1988/89, with the first of three years of additional grant funding, totalling $19.9 million, from the Atlantic Canada Opportunities Agency (ACOA), of which approximately half would be invested in capital renewal and half in programs and projects [151,152].

---

[28] Dr. Boorman had been with RPC since February 1966, and had served as head of the Mineral Development and Processing Department since 1978.

Figure 4.1. Dr. Roy S. Boorman, RPC's second permanent Executive Director, took office in 1983 and served until 1991. RPC photo.

In terms of governance, Mr. Cox stepped-down as Chairman of the Council in 1986 (but remained a Member of Council for five more years, until 1991). In his place, Dr. Knut Grotterod[29], was appointed RPC's third Chairman in November 1986 [153] (Figure 4.2). Dr. Knut Grotterod would go on to become RPC's longest serving Chair (at 22 years, see Appendix 8.2).

---

[29] At the time, he was President of Fraser Inc.

Figure 4.2. Dr. Knut Grotterod, RPC's third Chairman. RPC photo.

The RPC Act was amended from time to time over the years. Some of the more significant were the 1992 amendments to widen the Council's mandate to include more industrial research, production, management and worker retraining methods, and to increase the size of the Board to a maximum of 13 [154].

Figure 4.3. 1985-86 Marked the beginning of more colourful RPC Annual Reports (see Figure 2.5 for comparison).

Two more facilities expansions were constructed during *The Commercial Years*. An 8,000 sq. ft. RPC administrative office wing (Building #6 in Figure 4.4) opened in 1988 [155]. A major initiative enabled by the ACOA funding was the construction of an additional 18,000 sq. ft. building for RPC. This building (Building #7 in Figure 4.4), was to enable state-of-the-art laboratory and instrumentation capabilities in food, aquaculture, chemistry and biotechnology [156,157]. Particularly significant was the dedication of an entire floor in the new building for analytical chemistry services, which would become a core pillar of RPC's future business to the present day. It was opened on May 16, 1991 [89,158,159].

Figure 4.4. An RPC administrative office wing (Building #6, extreme right) was opened in 1988, and an additional laboratory building (Building #7, left of centre), housing new laboratories for food, aquaculture, chemistry and biotechnology, was opened in 1991. RPC photo.

In 1985/86, as part of a decentralization program, several regional RPC offices were opened [160]. The concept was to create a technology transfer network that could better promote close contact with industrial firms throughout the province. This enabled RPC to more conveniently offer the Technical Information Service, IRAP Program[30], and a window into all other RPC services, making them more accessible to entrepreneurs and small businesses. Regional offices were opened in Edmundston and Newcastle[31] in 1985, and in Moncton[32] and Saint John in 1986 [115] (Figure 4.5). Additional offices in Shippagan[33] and Bathurst were opened in 1989/90 [157].

---

[30] For which NRC provided additional funding.

[31] The Edmundston and Newcastle offices were closed in 1992/93.

[32] The Moncton office was closed in 2000/01.

[33] The Shippagan office moved to Caraquet in 1994/95, then from Caraquet to Grand Falls in 1997/98, then to Edmundston in 1998/99.

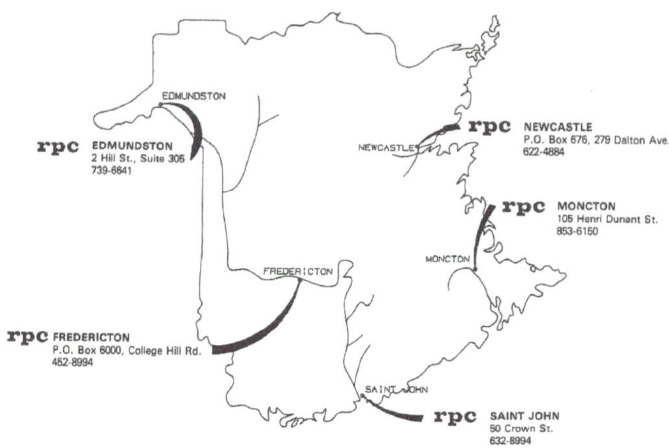

Figure 4.5. 1987 illustration of RPC's extended technology transfer reach via its central (Fredericton) and regional offices [115]. Additional offices in Shippagan and Bathurst were opened in 1989/90 [157].

Many RPC programs continued from the previous era, including the following.

Metallurgy and Non-Destructive Testing. Over the next few years, the metallurgical capabilities applied in support of the Point Lepreau nuclear power station, described in Chapter 3, were continued (Figure 4.6) but also provided to out-of-province clients including the Bruce and Darlington Nuclear Generating Stations in Ontario and the CANDU Owners Group generally [151]. In 1984/85 the Department of National Defense became a major client, with much of the work being in metallurgy (involving, for example, the fracture of plastics, underwater inspection of naval ship hulls, and the use of focused-beam ultrasonics to inspect defects in carbon fibre composites), particularly for the Canadian Patrol Frigate Program[34], beginning in 1988 [149,151,161,162].

---

[34] This program included an RPC training program for Royal Canadian Navy Officers, to give them *"a fundamental understanding of the metallurgical processes involved in fabricating frigates,"* to help them in their work overseeing and auditing the work being done by Saint John Shipbuilding Ltd. and MIL Davie Ltd. for the navy [161,162].

Figure 4.6. RPC's Johann Grimm conducting a fibre-optic examination of Point Lepreau pressure tubes. RPC photo.

Development of the SRL Electrowinning Process. Construction was completed and operations began at the $18.75 million pilot plant built in Chatham to test RPC's Sulphation-Roast-Leach process and operated by Enhanced Recovery Systems Ltd., an RPC subsidiary [163]. The plant was successfully started up and at year-end arrangements were being made for continuous test runs at a rate of 10 tons per day. The plant successfully demonstrated that the process would work in a commercial-scale operation. It wound-up in late 1986 [164].

Development of the Fluidized-Bed Coal-Oil Shale Process. RPC's high-sulphur-coal plus oil shale circulating-fluidized-bed combustor (CFBC) technology was pilot plant tested for a year, then advanced to be test/demonstrated in a $36 million project by NB Electric Power, also at Chatham [165]. Meanwhile, work continued on other fluidized-bed roasting applications as well, such as for peat, and by 1990, RPC had become a centre of excellence in this field [157].

Aquaculture. RPC also developed, for the aquaculture industry, a simple-to-use diagnostic kit for the rapid, inexpensive detection of diseases such as Bacterial Kidney Disease (BKD) and Furunculosis, in Atlantic and Pacific salmon (and trout) [166]. Named the *Agglutitest* diagnostic kit (Figure 4.7), it was licensed to Valox Ltd. in February 1991, for commercial production in Fredericton and sales worldwide [166-168]. This work led RPC researchers to help with 'The Salmon Crisis,' and to investigate the possibility of using live vaccine to protect salmon. As it turned out, RPC would play a crucial role.

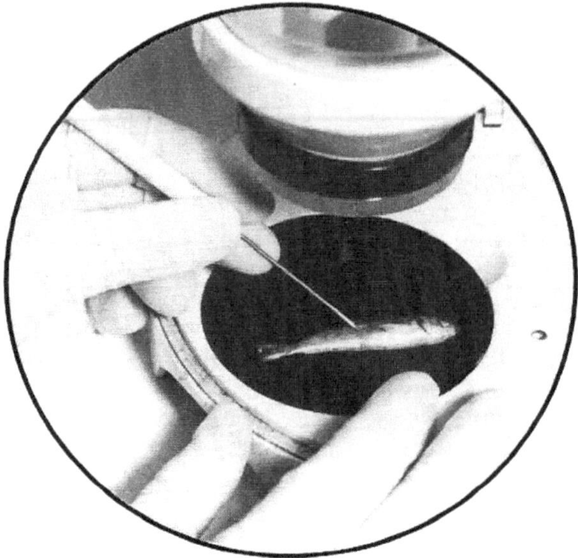

Figure 4.7. An RPC microbiologist testing an Atlantic Salmon for disease using the *Agglutitest* diagnostic kit, which involves antibody-linked latex-bead-technology. RPC photo, 1990 [167].

The Salmon Crisis. Dubbed *"One of New Brunswick's most spectacular 'bad news' stories of 1997"* [169], an unknown virus had been causing Haemorrhagic Kidney Syndrome (HKS) and killing Bay of Fundy salmon. As a result, the provincial Department had to order the slaughter of all infected salmon in the fish farms, resulting in approximately $20 million-worth of losses to 19 salmon farms in southwest New Brunswick [170]. RPC scientists were able to identify the virus that was causing the HKS and the deaths: Infectious Salmon Anaemia (ISA, Figure 4.8) [171].

Armed with the results of RPC's work, and learning from the experience of the Europeans in dealing with the disease, industry and government were able to take steps, including reducing the amount of effluent from fish-processing plants, to prevent a recurrence of the disease [170]. The impact of this work was saving a $120 million-per-year industry in New Brunswick. A lengthy story in the *New Brunswick Telegraph Journal* titled *"The Hunt for the Phantom Fish-Killer. How a Mysterious Virus Exposed Mistrust and Ill-Preparedness Among Fish Farmers, the Province and Scientists and Nearly Crippled a $120-million-a-year Industry,"* provides some of the inside story, including the competition among labs to to be the first to correctly identify the problem [172].

Scientists Isolate Virus Killing Atlantic Salmon

By SANDY MORGAN
For The Daily Gleaner

A virus first reported in the Norwegian aquaculture industry in 1984 has been identified as the killer of Bay of Fundy salmon.

Infectious Salmon Anaemia (ISA) was isolated by scientists at the Fredericton Research and Productivity Council last week. It causes Haemorrhagic Kidney Syndrome (HKS), which has affected 19 salmon farms in southwest New Brunswick, causing more than $10 million in losses. Having isolated the ISA virus, scientists confirmed their findings with the Central Veterinary Laboratory in Oslo, Norway.

John Kershaw, executive director of aquaculture with the Department of Fisheries and Aquaculture, said they have sought and are receiving vast amounts of information about ISA from Norway and are now attempting to put together a plan to manage the disease in the province.

"In Norway, as far as we understand, they have controlled the disease through measures like improved hygiene on the farms — by making sure there is no blood water allowed to contaminate the farm site, by removing mortalities as quickly as possible and by ensuring that all processing plants treat their effluent," said Kershaw.

Because these control measures reduced infected farms to four or five last year from 98 farms in 1990, Norway hasn't considered developing an ISA vaccine.

"In Norway, if ISA is identified on your farm, you must remove the fish out of the system immediately. They order you to slaughter and get them to market," said Kershaw.

He said he does not yet know, if they slaughter on a cage-by-cage basis or on a zone basis. The federal and provincial gov-

the disease.

The identification of the disease is just the first step in establishing a proper protocol to bring the disease under control.

"While this is an important finding, further research will need to be undertaken by RPC and other laboratories over the next few weeks to corroborate the test results," said Gay.

ISA suppresses the immune system in salmon, making the fish weak and vulnerable.

New Brunswick has 83 licensed farms and almost every one has reported some type of HKS case since the mysterious illness surfaced last fall. Between 16 and 18 of the farms have been seriously affected.

Some salmon farmers say be-

FOCUS

The hunt for the phantom fish-killer

How a mysterious virus exposed mistrust and ill-preparedness among fish farmers, the province and scientists and nearly crippled a $120-million-a-year industry

By Dave Young

BY THE TIME veterinarian John O'Halloran took the ferry over to Grand Manan and then headed to Seal Cove on a Friday early in the summer of 1996 the salmon farmer was already in the process of harvesting fish.

The call had come a day or so before about higher than usual deaths – or morts as they're called – in one of his cages. Usually, in a cage of 15,000 fish, about five to 10 fish die each week. This week that had jumped to more than 50. The salmon were already market ready and the farm was about to ship them to the fish plant.

A worker dipped a net into the cage and gathered up a few of the weaker fish.

N.B.-developed salmon vaccine promises to provide boost to worldwide aquaculture

BY ROGER LEBLANC
Telegraph-Journal

Fish farmers around the world have bigger and healthier salmon today thanks to a Fredericton research laboratory's discovery.

Salmon bacterial kidney disease is a big problem for aquaculture operations. Steve Griffiths, head of the molecular fish that have a very low food-to-weight-gain conversion ratio," he explained. "The worst affected fish are the ones nosing around in the pen at the surface, but there are a lot more infected that you know nothing about."

It's one of

'Hopefully it will save the industry millions of dollars.'

salmon in their pens. It seems the East Coast variety is more resistant to BKD. But there's been a constant worry about farmed Atlantic salmon escaping their cages and wreaking havoc on wild Pacific salmon.

Chile and here in New Brunswick.

Dr. Griffiths was initially looking for a way to better diagnose BKD. What he found, quite by chance, was another bacteria he's named *Arthrobacter davidanieli* after his young son, David Daniel Griffiths.

This bacteria, now known commercially as Renogen, is found naturally in the environ-

Figure 4.8. Headlines announcing RPC scientists' 1997 discovery of the virus responsible for $10 million-worth of salmon losses (top, [171]), the story of the race to identify (middle, [172]), and RPC scientist's discovery of a vaccine to prevent the disease (lower, [173]).

Additional protection was developed in subsequent work, through which RPC's Dr. Steve Griffiths and his team discovered a naturally-occurring bacteria, *Arthrobacter davidanieli*[35], that has a similar makeup to BKD but when injected into fish helps them to develop an immune response [173]. The vaccine was licensed to Novartis for sales under its commercial name, Renogen®, and it is currently registered to and manufactured by Elanco for sale in Canada and worldwide. According to RPC's Dr. Ben Forward, *"This vaccine has helped to provide a key Fish Health tool to protect fish from acquiring a disease that can result in significant economic losses to farmers"* [174].

Before leaving this topic, there was another chapter in RPC's applications of science in aid of the fisheries and aquaculture sector: the application of genetics to breed improvement. This work began in 2000 [175], and involved using DNA genotyping to identify desirable traits in fish, such as salmon, and to apply that knowledge to the breeding of the fish (Figure 4.9). In a media interview, RPC's Dr. Rachael Ritchie explained that *"[i]solating genetic markers for positive traits… allows farmers to breed the strongest, healthiest fish possible"* [176]. This provides the industry with a faster, more accurate, and more selective tool for breed improvement than the older methods of visual identification, physical tagging, and/or tagging. As noted in a media article with the eye-catching headline *'CSI' Science Helps Improve Genetic Diversity of Fish*, "[b]y maintaining and enhancing genetic diversity of broodstock, fish are healthier and grow bigger and stronger" [177]. In addition, genotyping can be used for 'traceability,' which is the ability to trace the origin of any particular fish, from the market, to the the processing plant, to the fish farm [177]. Traceability of a fish product can be particularly important, even crucial, to the international fish marketplace.

While much attention has been paid to salmon farming, similar advances have been made in the aquaculture of other fish, such as haddock. An example of an advance that was developed initially for haddock aquaculture is the RPC discovery of probiotic bacteria that can be introduced to newly hatched fish, enabling them to better tolerate otherwise harmful bacteria and experience lower rates of mortality (Figure 4.9) [178]. According to RPC's Dr. Ben Forward, *"The development of probiotic bacteria for use in hatchery culture of marine finfish and shellfish has delivered a solution to alleviate a critical bottleneck in the production cycle while providing an alternative to more costly and less sustainable alternatives"* [174].

---

[35] Named after Dr. Griffith's son, David Daniel Griffiths.

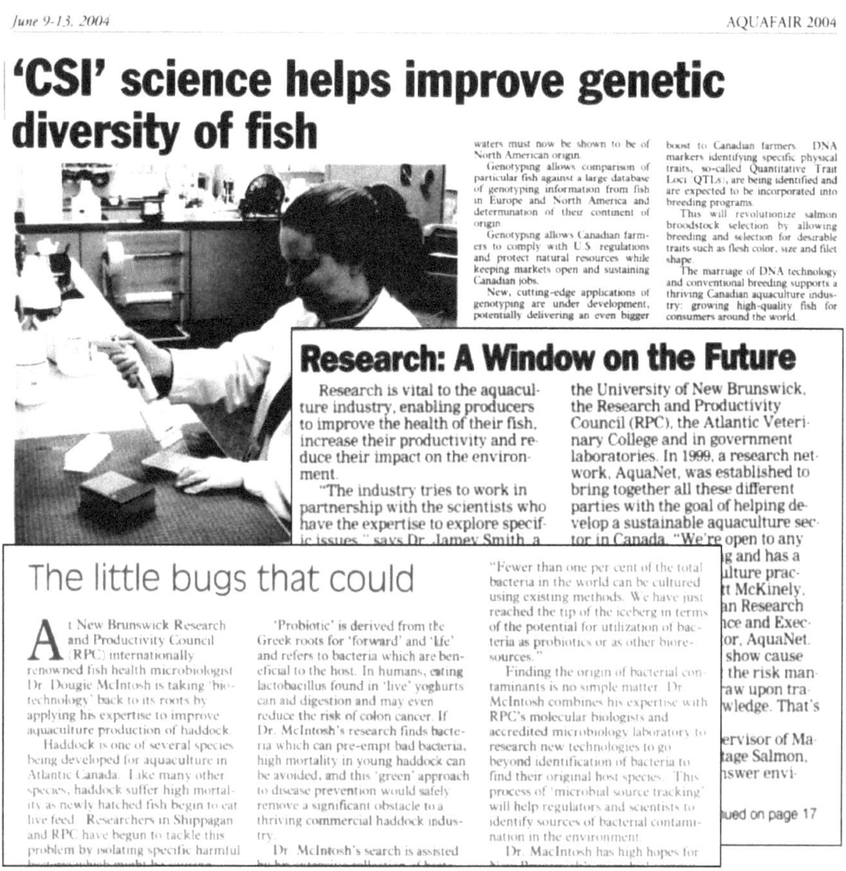

'CSI' science helps improve genetic diversity of fish

waters must now be shown to be of North American origin.

Genotyping allows comparison of particular fish against a large database of genotyping information from fish in Europe and North America and determination of their continent of origin.

Genotyping allows Canadian farmers to comply with U.S. regulations and protect natural resources while keeping markets open and sustaining Canadian jobs.

New, cutting-edge applications of genotyping are under development, potentially delivering an even bigger boost to Canadian farmers. DNA markers identifying specific physical traits, so-called Quantitative Trait Loci (QTLs), are being identified and are expected to be incorporated into breeding programs.

This will revolutionize salmon broodstock selection by allowing breeding and selection for desirable traits such as flesh color, size and filet shape.

The marriage of DNA technology and conventional breeding supports a thriving Canadian aquaculture industry: growing high-quality fish for consumers around the world.

Research: A Window on the Future

Research is vital to the aquaculture industry, enabling producers to improve the health of their fish, increase their productivity and reduce their impact on the environment.

"The industry tries to work in partnership with the scientists who have the expertise to explore specific issues," says Dr. Jamey Smith, a the University of New Brunswick, the Research and Productivity Council (RPC), the Atlantic Veterinary College and in government laboratories. In 1999, a research network, AquaNet, was established to bring together all these different parties with the goal of helping develop a sustainable aquaculture sector in Canada. "We're open to any g and has a ulture prac- t McKinely, an Research ce and Executor, AquaNet. show cause the risk management draw upon trawledge. That's ervisor of Matage Salmon, answer envi-

The little bugs that could

At New Brunswick Research and Productivity Council (RPC) internationally renowned fish health microbiologist Dr. Dougie McIntosh is taking 'biotechnology' back to its roots by applying his expertise to improve aquaculture production of haddock.

Haddock is one of several species being developed for aquaculture in Atlantic Canada. Like many other species, haddock suffer high mortality as newly hatched fish begin to eat live feed. Researchers in Shippagan and RPC have begun to tackle this problem by isolating specific harmful

'Probiotic' is derived from the Greek roots for 'forward' and 'life' and refers to bacteria which are beneficial to the host. In humans, eating lactobacillus found in 'live' yoghurts can aid digestion and may even reduce the risk of colon cancer. If Dr. McIntosh's research finds bacteria which can pre-empt bad bacteria, high mortality in young haddock can be avoided, and this 'green' approach to disease prevention would safely remove a significant obstacle to a thriving commercial haddock industry.

Dr. McIntosh's search is assisted

"Fewer than one per cent of the total bacteria in the world can be cultured using existing methods. We have just reached the tip of the iceberg in terms of the potential for utilization of bacteria as probiotics or as other bioresources."

Finding the origin of bacterial contaminants is no simple matter. Dr. McIntosh combines his expertise with RPC's molecular biologists and accredited microbiology laboratory to research new technologies to go beyond identification of bacteria to find their original host species. This process of 'microbial source tracking' will help regulators and scientists to identify sources of bacterial contamination in the environment.

Dr. MacIntosh has high hopes for

ued on page 17

Figure 4.9. Headlines announcing RPC scientists' discoveries and applications of genetic tools to aid the New Brunswick Aquaculture Industry, *circa* 2004. From references [177] (top), [176] (middle-right), and [178] (lower-left).

Several new programs also made their appearance in this era, including the Manufacturing Technology Centre and a new program in Electrochemistry:

Manufacturing Technology Centre (MTC). This centre was launched in 1985 with provincial funding support from the Department of Regional Industrial Expansion. RPC's MTC was part of a larger initiative that was based at UNB (created in 1983), and which had additional centres at RPC, Université de Moncton, and three community colleges [179].

One of the key enablers of MTC was a technology transfer agreement between the federal and provincial governments and McDonnell Douglas

Corp. (of St. Louis, Missouri), under which the latter company's world-leading Unigraphics CAD/CAM technology was licensed[36] [180,181]. This would become a great example of a successful government industrial-benefit program in New Brunswick because, armed with the new technology, RPC's MTC was created to transfer CAD/CAM (Computer-Aided Design, Computer-Aided Manufacturing) technology to New Brunswick companies. 1985/86 was the RPC-MTC's first full year of operation (Figure 4.10) [160].

With RPC's help, many companies were able to learn how to use the new tools to improve their ability to manufacture products requiring *"rigid specification to stringent quality assurance"* [182]. By the following year, major clients such as SPAR Aerospace had been engaged, and it had become possible to distribute PC-based CAD systems directly into industrial clients' facilities [115].

In 1989, a new computer network enabled MTC to be connected to CADMI Microelectronics Inc. (Centre for the Application and Development of Microelectronic Inc., a UNB not-for-profit company) [183], and a study of industry needs was conducted to determine whether the two organizations should be merged [184]. By June 1990, MTC and CADMI were working in partnership as a 'network of centres and facilities,' and co-branding their alliance as MTC/CADMI [185].

To extend RPC's reach, an office with five CAD/CAM terminals was located in Edmundston to help the northwest region of New Brunswick to apply advanced manufacturing technology.

In addition to the host computer system supporting the CAD/CAM activities, a major addition was an advanced numerically-controlled machine tool for technology demonstration. Among the first industry projects were two in robotics development [180]. The first such project involved the design and manufacture of a prototype system for loading silicon wafers into an oxidation furnace. The system was ultimately commercialized by RPC's client, a supplier of processing equipment to the semiconductor industry. The second project involved a novel method of removing wire from pulp bales. RPC developed the robotic concept, identified a local manufacturer to build, construct and market the machine and undertook to design and construct a prototype machine.

Among the many other examples of economic impact resulting from RPC's help in this area was a woodworking firm in northwestern N.B., for whom RPC helped develop a series of multi-shaped utility boards, which enabled them to widen their product base [186].

---

[36] Unigraphics was famous for its 3D wire frame modelling software, which was a significant advance over the standard (at the time) 2D-CAD systems. Both were forerunners to today's 'solid' models.

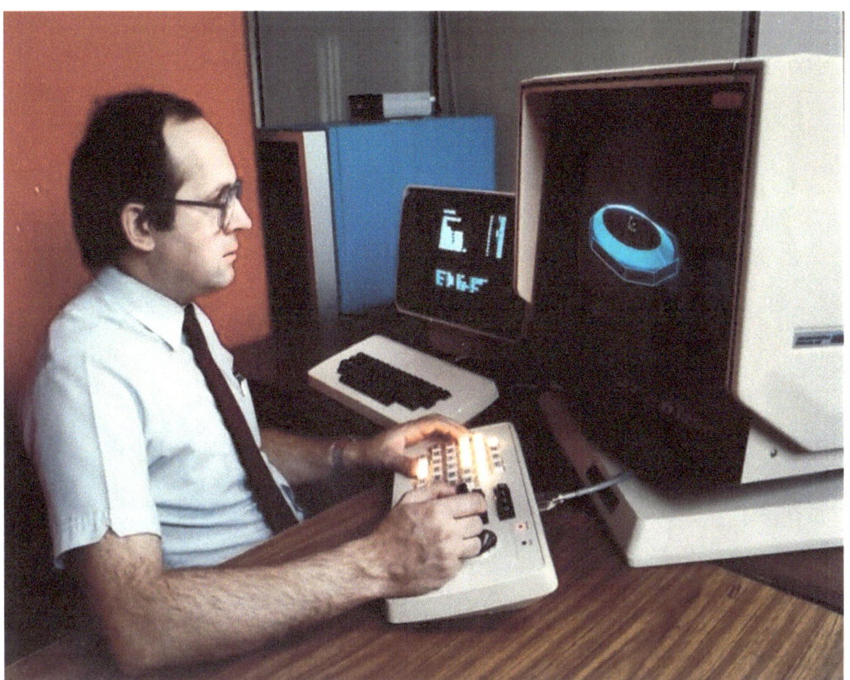

Figure 4.10. RPC's Michal Kuczewski and a Unigraphics CAD/CAM workstation at the Manufacturing Technology Centre *circa* 1986. RPC photo.

Electrochemistry Program. In 1978, RPC began working in battery technologies. Before long, this turned into a dedicated electrochemistry group that developed capabilities in: electro-organic synthesis, high-energy battery technologies, preventative corrosion analysis, and battery testing. Important advances had been made by 1984 [187]. For example, contract R&D on high energy density battery systems for the Canadian Department of National Defence enabled the group to develop sufficient expertise to successfully land, in 1984, an industry-sponsored, multi-year contract to develop a lithium rechargeable 'C-cell' battery prototype [188]. The New Brunswick Telephone Co. became a major client for this work. Additional work focused on improving battery cycle life to over 1,000 cycles and extending batteries' low temperature performance to -40 C.

By 1986/87, the group had developed a hermetically-sealed, rechargeable 'C-cell' prototype, ready for pilot-plant production and commercialization (Figures 4.11, 4.12). Meanwhile laboratory-scale work was underway to develop lithium 'AA-cell' prototypes. Several aspects of the technology development hinged on RPC inventions, like a new kind of battery electrolyte and a new kind of lithium-based, anode-alloy material called linode (e.g., Canadian Patent 1,302,491 and U.S. Patent 4,888,258; see Appendix 8.8).

Figure 4.11. RPC's Kevin O'Neill conducting glove-box work in lithium-battery electrochemistry. RPC photo, 1987 [115].

Figure 4.12. An RPC prototype lithium-rechargeable battery *circa* 1987 (Left, [189]) and one of the first commercial-product versions (Right, [89]). RPC photos.

RPC's work on rechargeable lithium batteries had some unintended consequences. In July 1990, for example, a fire broke out requiring evacuation of the buildings and some care in extinguishment because it was a 'lithium fire.' In this case, propylene carbonate solvent was being used to treat a coil of lithium metal that was being used in RPC's lithium-battery development work. Unfortunately, the solvent (being hydroscopic) had absorbed some water [190]. When contacted by water, lithium metal spontaneously bursts into flame, so this was not the kind of fire that could be put out by adding more water to it! According to RPC staff [190], three Class-D fire extinguishers[37] were used on the fire but, according to RPC's Dr. John Macaulay, "*The fire was still going and there was a cloud of graphite dust in the air and so at that point everybody cleared out of the lab*" [190]. The fire department was able to put the fire out. Then, with the assistance of Canadian Forces Base Gagetown, who provided 25 gallons of mineral oil, RPC staff and the fire department covered the lithium with mineral oil (Figure 4.13), while the containment barrels were sprayed with carbon dioxide to keep them cool [191].

Figure 4.13. RPC's Tim Doherty and a Fredericton firefighter cover lithium metal (in the containment barrels shown) with mineral oil in 1990 [191].

---

[37] Class-D fire extinguishers (for flammable metals) contain dry powders such as powdered graphite or granular sodium chloride.

Contract revenues continued to grow in this era, and by 1985/86 reached approximately 91% of total revenue [160]. It would keep on rising, although more slowly, through to the end of this era (see Figure 4.38, below). Such strong and sustained contract revenues were supported, in part, by strong marketing and advertising, and also by positive media reporting. Some examples of RPC advertising in the mid-1980s are shown in Figure 4.14.

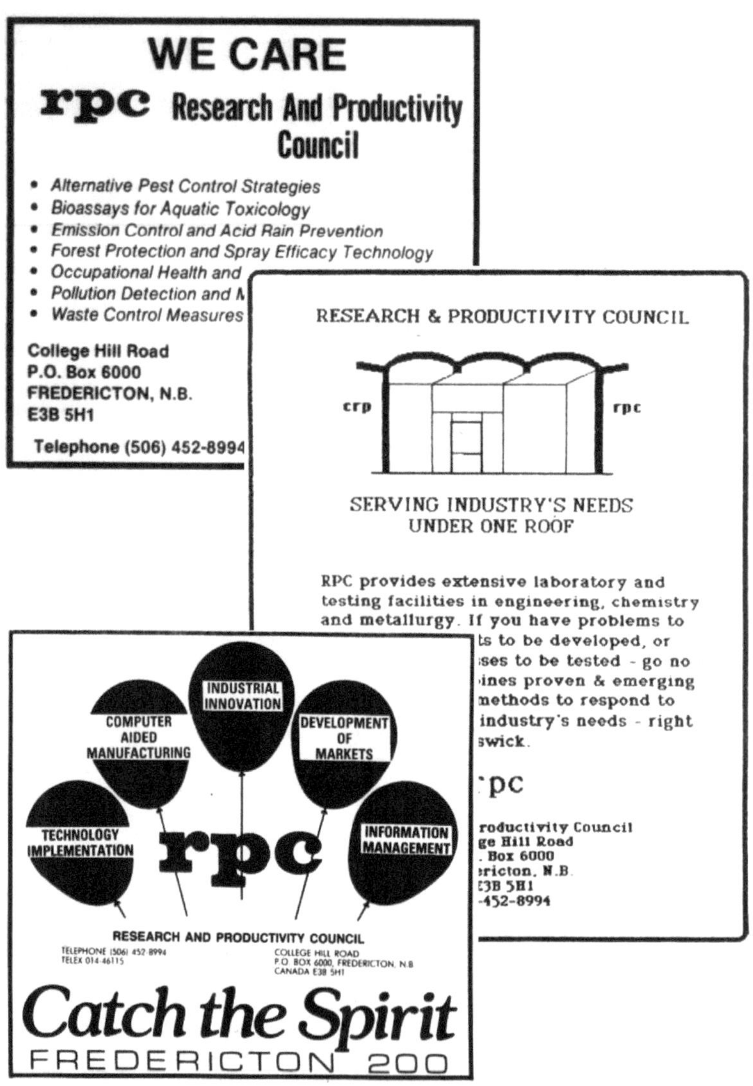

Figure 4.14. RPC media ads from the *Saint John Telegraph Journal*, June 5, 1984 (upper left) and the *Fredericton Daily Gleaner*, Feb. 14, 1986 (centre right), and April 25, 1985 (lower left).

Of course, RPC's strong reputation was a huge asset contributing to public and private awareness, support, and business. The 1985/86 Annual Report [160] provides some examples of RPC's contributions to industrial innovation and business competitiveness, as summarized by RPC's Council Chairman, Mr. Cox:

"… the development by RPC engineers of field techniques for determining the microstructure of metallic components and for predicting the remaining service life. On-site application, typically on an emergency basis during planned or unplanned shutdowns, has provided unambiguous answers to critical and urgent questions at several industrial plants… Problems which threatened to delay production start-ups at pulp and paper mills, utility boilers and oil refineries were resolved in hours when days were required previously to answer those same questions."

Detailed, positive media coverage of RPC in the 1980s continued to be the norm, with the media running stories with titles like "RPC Meeting Challenges of Today" [192], and "Council Benefits Many in Province" [193] (see also the headlines in Figure 4.15). For example, in a June 5, 1984 article titled "It's a Technology Centre for N.B. Industry," the *Telegraph Journal* reported [194]:

"Over a period of time there has been a return of more than 10 times the money spent in terms of process innovation, manufacturing process improvement, quality assurance maintained or jobs created. It has built up an enviable record both in the province and elsewhere in Canada for know-how and do-how."

It was a strong testament to RPC's reputation with all stakeholders that media coverage of RPC continued to be positive when its name was being mentioned, virtually daily, during New Brunswick's 'Rancid Tuna Scandal' of 1985. The latter topic, also dubbed 'Tunagate,' is addressed next.

The Telegraph Journal and The Evening Times-Globe, Tuesday, June 5, 1984   3

It's A Technology Centre For N.B. Industry

The New Brunswick Research and Productivity Council which has a staff of 120 and a contract revenue of $5 million is a low profile, result-orientated research and development centre which the province can be proud of. It earns most of its own keep through contract research and the return on investment to the province has been large. Over a period of time there has been a return of more than 10 times the money spent in terms of process innovation, manufacturing process improved, quality assurance maintained or jobs created. It has built up an enviable record both in the province and elsewhere in Canada for know-how and do-how. RPC is in the forefront of resource and technology development work for New Brunswick as well as doing a large amount of service work for government and industry.

The resource development work is aimed at exploiting the unused mineral and energy wealth of the province. A good example is the work, which started more than 10 years ago, to develop the deposits of lead, zinc and copper of the Bathurst area which could not be profitably exploited. The RPC sulphation roast leach electrowining (SRLE) process will now make this possible. A pilot plant demonstration of the process is to take place at Chatham. The federal government is providing $15 million and the provincial $3.75 under a Canada-New Brunswick agreement signed in 1983. Other firsts for RPC have included showing how the oil shales of Albert County can be utilized to reduce our dependence on imported oil. Work is now being conducted at RPC to test retorting processes for recovering oil from the shale. RPC, financed by the Department of Natural Resources, made the initial technology breakthrough which has resulted in the $36 million demonstration for NBEPC at Chatham in the co-combustion of oil shale and coal in a fluidized bed to generate electricity.

An important part of technology development has involved forest protection. The behaviour modifying chemicals (pheromones) being developed offer an alternative to present pesticides for the control of forest insect pests. In addition, RPC is coordinating a multi disciplinary research group of scientists from the National Research Council, Canadian Forest Services and the New Brunswick universities to improve the efficiency of aerial spray techniques.

Technology development with the object of helping existing industries or starting new businesses includes RPC's assistance in the application of computer aided design and computer aided manufacturing (CAD/CAM) so that industries can compete for orders to supply manufactured parts to clients elsewhere. RPC assists firms in product design, manufacturing, process planning, part programming and prototype development.

Much day-to-day service is also provided to government and industry. On behalf of the province it is contracted to administer programs under the Conservation and Renewable Energy Demonstration Agreement and, through the provincial Energy Secretariat, it is responsible for the National Energy Audit Program. This conservation program has been particularly successful. RPC has developed the microcomputer energy analysis method or "suitcase audit" which has allowed trained staff, equipped with a package of briefcase size, to identify some $10 million of potential energy saving for 900 clients in New Brunswick.

On behalf of the National Research Council and the New Brunswick Department of Commerce and Development, RPC administers the Industrial Research Assistance Programs (technical information and industrial engineering) and the Small Scale Industry Financial Assistance Program. Many industries and individual businesses have benefited from these programs.

There are many other activities at RPC such as mechanical technology, metallurgical engineering, food science and analytical chemistry. Arrangements can be made with the executive director to visit the laboratories and workshops in Fredericton.

4a   The Daily Gleaner, Thursday, February 21, 1985

RPC Provides Technological Excellence

Businesses and entrepreneurs use the Research and Productivity Council (RPC) because it provides technological excellence with quick and confidential assistance in solving problems to help meet company objectives. Many companies and corporations who want to benefit from new avenues of research and development or need current work extended, come to RPC because there are over 100 scientists, engineers, technicians and support staff ready to serve and millions of dollars of special equipment to be used.

Both staff and resources can be allocated to specific industrial research projects as the occasion warrants. Dedicated personnel provide the additional technical skills needed to develop and test workable commercial processes or to identify opportunities for cost cutting and to provide the many other technical services required to maintain business profitability. RPC keeps up to date with technology relevant to New Brunswick and what opportunities there are for its application.

Since 1981, almost $1 million has been spent, through the industrial research assistance program (IRAP) of the National Research Council of Canada, in solving nearly 200 engineering and technical problems encountered by small and medium sized manufacturers in New Brunswick. In the course of this work RPC has built a machine to form the ribbon shape in candy. It developed new areas of waste wood utilisation which have potential industrial spin-off, and it has demonstrated the financial attractiveness in developing fish and shellfish farming off the Northumberland Strait as an income supplement for seasonally employed fishermen.

Manufacturing technology has an increasingly high profile at RPC. RPC assists industry apply computer aided design and manufacturing (CAD/CAM) to production problems. The means of manufacturing to rigid specifications with stringent quality assurance are improved considerably by use of these advanced technologies play an important role in establishing new enterprises by assisting firms in product design, manufacturing, process planning, parts programming and in prototype development.

Not only does RPC respond to industries' needs, but it takes the initiative in development opportunities. The engineers and scientists generate ideas based on familiarity with current and new technology which, coupled with identifying market opportunities, is matched with an industrial need such as under utilized natural resources or the need to economically dispose of a waste product.

Funding is sought for this mission oriented research if it cannot be financed in house. More money is needed as development proceeds until finally the project has reached the stage that venture capital can be attracted to put a product or process on the market.

An outstanding example is the mineral processing innovation for the production of zinc, lead, copper and silver now being demonstrated at Chatham by Enhanced Recovery Systems Ltd., a subsidary of RPC. Another is the pioneer work done on oil shale utilisation which has helped so much in obtaining the federal investment in the Chatham plant of the New Brunswick Electric Power Commission for the use of oil shales in electricity generation.

Other RPC-initiated work with promise of commercial spin-off deals with the manufacture of special rechargeable electical batteries and a method of disposing of hazardous wastes from industrial processes.

RPC engineers have been especially successful in helping industry. A small in-house project to determine by remote control methods, the flow of liquids and gases through pipes has allowed the development of an important tool for the nuclear industry. Being able to determine flow stoppage before they are an industrial problem can save millions of dollars in preventing down time.

Advanced technologies such as these have been applied to solving problems in a wide variety of industries. The savings to companies have been large and to this should be added the benefits of the multiplier effect. For instance help to a pulp mill ensuring its uninterrupted operation has important ramifications to those employed in the woods, which in turn affects retail trade and so on.

Figure 4.15. Positive headline news coverage of RPC activities in the *Saint John Telegraph Journal*, June 5, 1984 (upper) and the *Fredericton Daily Gleaner*, Feb. 21, 1985 (lower).

RPC and 'Tunagate,' the 'Rancid Tuna Scandal.' It began in early September 1985, when CBC's *The Fifth Estate* program reported that the federal fisheries minister had *"overruled his department's inspectors"* the previous April and *"allowed distribution of almost one million cans of tuna which earlier had been declared unfit for human consumption"* [195,196]. According to further media reporting, *"two levels of federal inspectors [had] rejected the cans, partly because tuna in them was rancid and decomposing"* but the minister had *"accepted recommendation by the Research and Productivity Council... which said the tuna was fit for human consumption"* [195,197].

Figure 4.16. Headline news coverage of the 1985 'rancid tuna scandal.' Clockwise from the top: *Moncton Times Transcript* [196], *Fredericton Daily Gleaner* [200], *Moncton Times Transcript* [202], *Saint John Telegraph Journal* [197].

The media raised questions over, among other things, whether the federal minister had bowed to pressure from New Brunswick Premier Richard Hatfield in making his decision, and whether RPC had been influenced in its work by instructions from the provincial government to *"save the company and the jobs of its employees,"* and whether the *"two levels of government should have not allowed themselves to be 'pushed around' by an American-owned company…"* [197,198]. The 'company' in this case, was Star-Kist, headquartered in St. Andrews, and producer of the two brands identified in the issue: Ocean-Maid and Bye-the-Sea. As time went on, it became known variously in the media as 'Tunagate' and the 'Rancid Tuna Scandal.'

RPC had conducted tests of the fish over a two-to-three-week period involving about 20 people in taste and smell tests [199] and also subjected them to detailed laboratory analysis [200]. According to Premier Hatfield, *"What RPC (the Research and Productivity Council) did was do a chemical analysis of it to try to determine decomposition. They (the government inspectors) did it with the senses rather than any kind of chemical test… I want to make the point that I have a great deal of confidence and respect for the integrity of the Research and Productivity Council. They have an international reputation. I think they have acted very properly and in a scientific way"* [201].

Subsequent testing by RPC on additional samples did find some lots of tuna that were recommended for rejection (approximately ten percent) [200,202], but by this time the minister had already made his ruling. Although there do not seem to have ever been any reported health problems, and the products in question were eventually withdrawn from most stores, the media and politicians focused much attention to questions of who had known what and when, and to future standards.

RPC's reputation remained intact during the two months of intense media coverage (Figure 4.16). In the words of Dalton Camp, in a Toronto Star article [203]:

"it was agreed that an independent and qualified 'third party' would be engaged to treat samples of the product according to federal standards and practice… The organization retained for these tests was the Research and Productivity Council of New Brunswick, whose integrity and competence were acknowledged by all parties to the dispute."

In the end, Premier Hatfield also weighed-in, announcing that *"The expertise of [RPC] has been used to improve inspection procedures at the Star-Kist tuna canning plant… and to enhance the objectivity of federal inspectors which in the past was doubtful"* [204].

Some of the evolution of RPC's principal program and service offerings can be seen by comparing the following 1987/88 summary [150] with those in Chapters 2 and 3:

- o ***Chemical and Biotechnical Services*** - Crop and forest protection, insect detection and surveys, spray efficacy, droplet deposition, aerial application techniques and spray drift (Figure 4.17), alternate pest control technology, development and manufacture of pheromone products, pesticide residue analysis, organic synthesis.
- o ***Electrochemistry*** - Lithium battery research and development, battery testing, electro-organic synthesis, advanced materials research.
- o ***Engineering Services*** - Machine design and fabrication, prototype development, measurement and calibration, vibration analysis, CAD/CAM applications, robotics, N/C programming, electronic design and fabrication, control systems, data acquisition systems, energy audit.
- o ***Engineering Materials*** - Nondestructive testing, inspection, quality assurance, training, failure analysis, materials testing, corrosion testing, welding procedures, fracture mechanics, stress relieving, helium leak detection, ultrasonic fluid flow metering, fibre optics.
- o ***Food, Fisheries and Aquaculture*** - Product and process development, quality control, trace analysis, microbiological examination, aquaculture systems development, gear and cage design, fish health services, environmental monitoring programs.
- o ***Industry Services*** - Industrial engineering, productivity improvement market research and planning, computer applications and training, feasibility studies, Industrial Research Assistance Program (IRAP) delivery in partnership with NRC.
- o ***Mineral Development and Processing*** - Resource characterization, process development process evaluation, mineralogical analysis, beneficiation strategies, exploration aids, waste management, analytical services, alternate fuel technology, combustion research and demonstration, fluidization technologies, pilot plant design and operation, aerial emission quality.
- o ***Other Services*** – The provision of comprehensive information services to New Brunswick industry, including standards retrieval, patent searching. and online database access to worldwide sources of technical and scientific information.

Of these, the chemical and biotechnical services, in particular, continued to grow, doubling in the following year (1989), driven by demand for trace organic analyses related to increased public and regulatory concerns around environmental contaminants [157].

Figure 4.17. Spray Technologist Chris Riley collecting samples to assess off-target insecticide-spray drift in an apple orchard in the Saint John River Valley. RPC photo, 1988 [150].

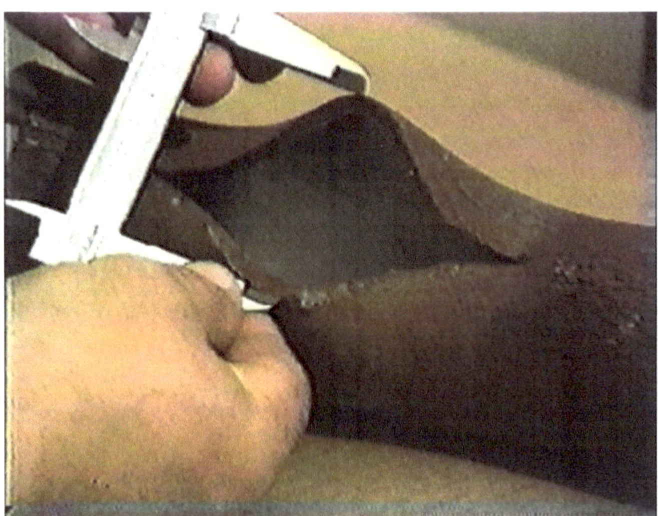

Figure 4.18. RPC metallurgy: investigating the cause of material failure in a burst pipe. RPC photo, 1986 [189].

Forest and Crop Protection. Advances were continuing to be made in such areas as behaviour-modifying pheromones as an alternative to conventional pesticides, and new techniques to improve the efficiency of aerial crop spraying. Both of these were primarily pursued in response to the Spruce Budworm Crisis.

Spruce Budworm Control. Forest insects are the biggest cause of tree mortality in Canada [205]. For New Brunswick, the greatest concern has been with the Spruce Budworm, a native moth that, as a caterpillar, can defoliate fir and spruce trees. Outbreaks of the Spruce Budworm in Eastern Canada have been cyclical, having occurred approximately every 30 to 40 years, and in epidemic proportions since at least the 1960s. Among other problems, severely infested spruce trees lose their value as saw logs and/or for making pulp and paper products.

RPC worked for many years (dating from 1975 [92]) on methods to combat the eastern spruce budworm (*Choristoneura fumiferana*) and the spruce budmoth[38] (*Zeiraphera canadensis*). One of the most common means of combatting Spruce Budworm outbreaks has been aerial spraying of insecticides, an area of RPC research since the beginning of the 1980s [123] (Figures 4.17 and 4.19). The pesticide approaches developed and refined ranged from conventional organic-chemical pesticides like fenitrothion, biological pesticides like *Bacillus thuringiensis* (Bt), and sex pheromones [206,207].

In the latter case, RPC scientists developed synthetic sex pheromones, identical to the scents produced naturally by female budmoths, as a natural alternative to traditional chemical pesticide spraying [92,124] (see the 1980s pheromone patents listed in Appendix 8.8). When delivered from controlled-release dispensers in white spruce plantations (Figure 4.20), the released pheromones confuse the male moths and prevent or reduce the amount of mating and therefore the numbers of fertile eggs being laid. A 1994 field test of the technology, carried out at a J.D. Irving Woodlands Ltd. white spruce plantation at Boston Brook, N.B., achieved an 80% reduction in viable egg production [206]. Further work involved the development of a microencapsulated formulation for use in large-scale spraying from aircraft.

---

[38] The spruce budmoth tends to feed on open-growing spruce trees rather than spruce trees in forests, which tend to be attacked by the spruce budworm.

Figure 4.19. Cessna Agtrucks mounted with rotary atomizers spray the Bt pesticide during field trials to determine the effects of post-spray weather on pesticide performance. RPC photo, 1988 [151].

Figure 4.20. Delivering synthetic sex pheromones from controlled-release dispensers in a white spruce plantation. RPC photo, 1996 [208].

RPC Lab Technician Mike Skinner attaches a Shin-Etsu rope dispenser at the test site.

In 1987, RPC celebrated 25 years of success *"in support of industries that are driving the New Brunswick economy"* [150] (Figures 4.15, 4.16). Among the celebrations was a 25th Anniversary Open House that was held on Friday, May 1 through Saturday May 2, 1987. This included displays and laboratory tours and demonstrations (Figures 4.17, 4.18). Among the highlights noted were:

o   Challenging and helping the mining and energy sectors to exploit new technology to achieve increased recoveries and greater resource utilization,

o   Contributing to the development and application of new technology for insect control to protect the quality of new and existing forest resources,

o   Supporting the fisheries industry with improvements in raw material handling practises and processing techniques for high quality herring-roe products,

o   Improving cost effectiveness in aquaculture by reducing the risk of disease outbreak and improving the nutritive value of fish feed formulations,

o   Assisting pulp and paper mills, chemical plants and power utilities through the provision of specialized inspections and testing to improve the reliability of operating equipment and reduce downtime, and

o   Helping New Brunswick's smaller manufacturers with a wide range of technical assistance such as product development, new manufacturing technology, and product testing.

Also in 1987/88, RPC launched an in-house R&D program that was financed from the increase in the provincial grant awarded at the beginning of the year. The objective was to enable staff to pursue innovative ideas sparked by their exposure to the resource and manufacturing sectors. The only requirement was that projects had to lead to client interest and sponsorship within one year thus making the program a potentially valuable stimulator of product or service opportunities for New Brunswick companies [150].

Figure 4.21. In 1987, RPC celebrated its 25th Anniversary [150].

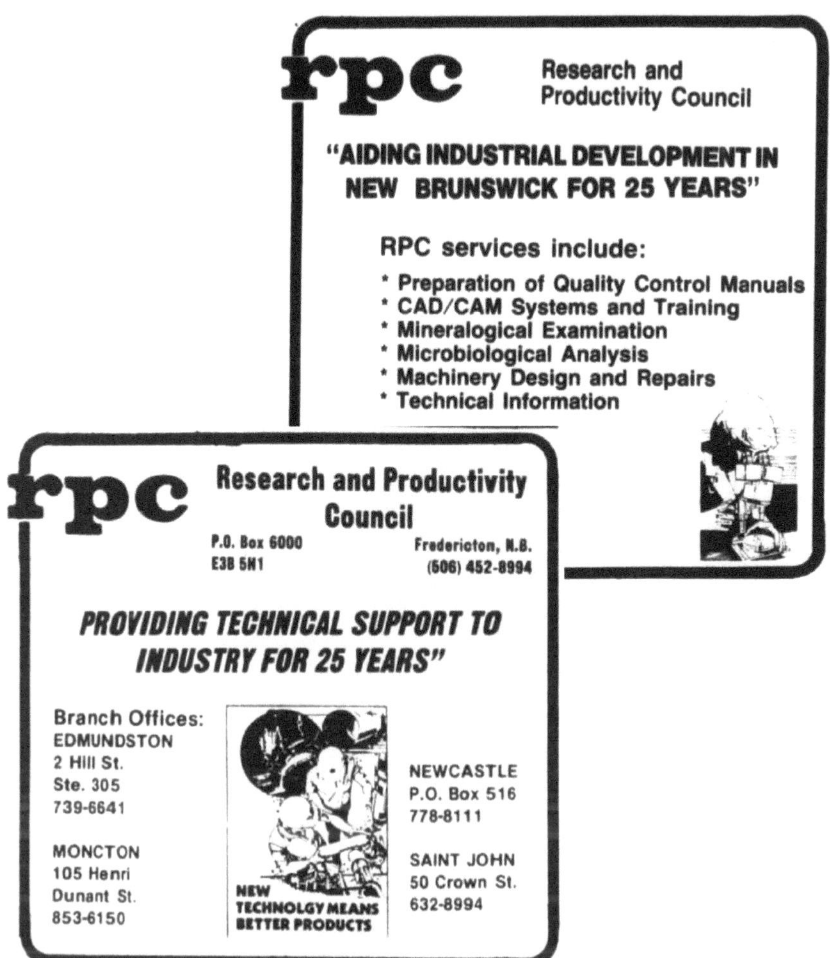

Figure 4.22. RPC 25th anniversary media ads from the *Fredericton Daily Gleaner*, Feb. 12, 1987 (upper right) and the *Saint John Telegraph Journal*, Feb. 26, 1987 (lower left).

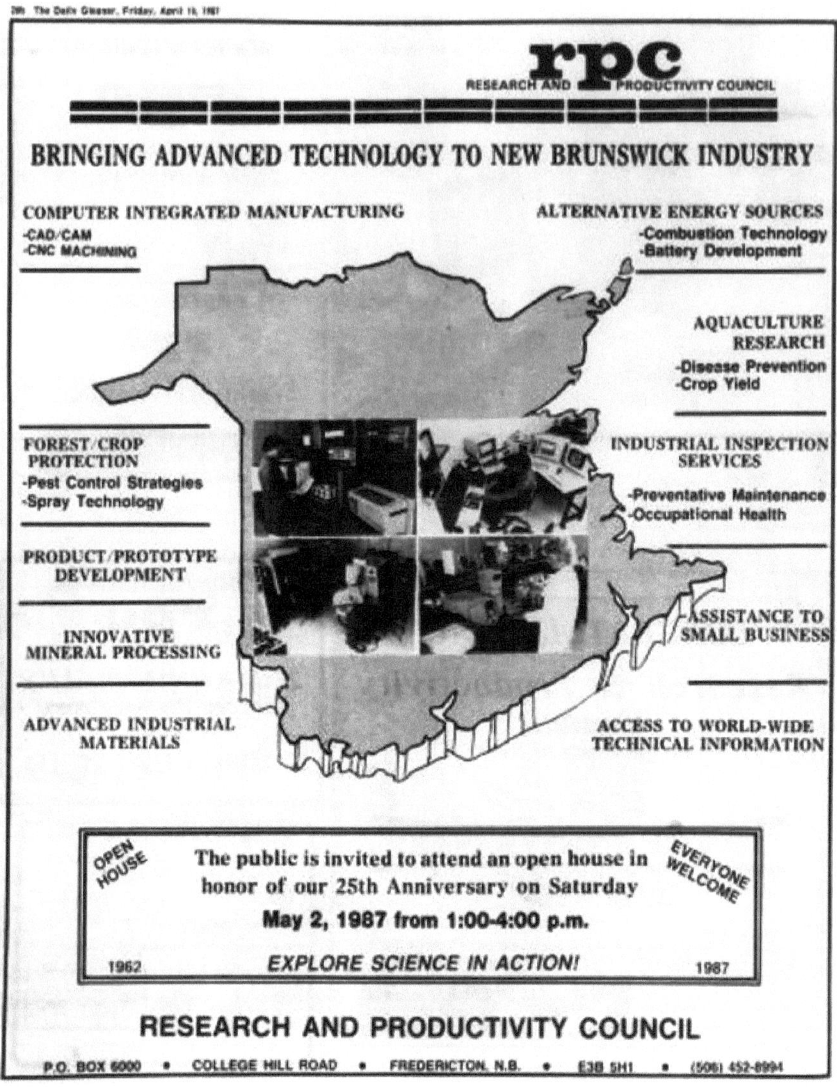

Figure 4.23. RPC 25th anniversary and Open House media ad from the *Fredericton Daily Gleaner*, April 10, 1987.

It wasn't all celebration in 1987, however. RPC's total revenue had just dropped for the first time in its corporate history (mostly due to decreased provincial government contracts) [115]. Although RPC would overcome the challenges that this presented, the road ahead would be bumpy.

Figure 4.24. RPC's Frank O'Connor demonstrates the operation of a CNC Vertical Machining Centre at RPC's 25th Anniversary Open House, May 1-2, 1987. RPC photo.

RPC continued to provide information services throughout *"The Commercial Years,"* both internally and externally, as the amount of technical information that could be made available to RPC staff and to industry clients continued to grow. By 1984, for example, the library had 10,000 volumes and 250 periodicals [209] and it kept on growing with each successive year. To increase timely dissemination of information, RPC launched several periodical publications including:

o *"**Inside Out**,"* a nominally bimonthly internal newsletter aimed at RPC employees and their families, and which was published from 1985 through 1995 (Figure 4.25), and

o *"**Alert,**"* a quarterly newsletter, was published from 1984 (Figure 4.26) through 1994, at which time it was made triannual and renamed *"**Visions,**"* which was published through to 1996. *Alert*, and later *Visions*, were sent to past and potential clients to update them on emerging technological developments and, of course, RPC's projects and expertise.

The first issue of *Alert* was published in 1984 [209] (Figure 4.26), with articles on RPC projects in candy production, wood waste utilization, cultured blue mussel introduction, energy efficiency audit services, and the nature of the IRAP services available through RPC.

Taste tests a delicious experience

This group sampled some new recipes at an RPC focus group taste test in July.

I n 1992, a notice went out that RPC's Consumer Product Specialist Heather Wilson was looking for RPC volunteers to taste test a lobster bisque. Those that volunteered received some insight into the consumer market research that Heather conducts on a daily basis for various clients.

Heather explained that she deals with two different types of clients: clients that have a real product and clients whose product is at the idea stage that hire RPC to determine if the idea can be developed into a product. Depending on which category the client fits, one of two different approaches is taken.

"For someone with an actual product, our job is to validate that all of the features on the product meet consumer demands," said Heather. "So, our first job is to identify what consumer group would be interested in buying this product and then seek those people out to ask them questions about whether or not this product would interest them and if the features in the product meet their needs."

If Heather learns that the features do not meet consumer needs, she identifies what changes are necessary to create a demand for the product. The consumer group is typically made up of individuals with one or several common traits such as age, lifestyle, nationality or disposable income.

"Since the second group does not have a product at this stage, we take their concept and develop it. The client often needs us to define what features the product should have, what price it should be, or where it should be sold and we determine those answers through market research," Heather added.

Initially, RPC's market research group contacts their 'key informants'; people in the industry that have some affiliation with the product. They start by asking questions to develop a sense of where the product fits. From that information, RPC staff decide whether telephone interviews or personal interviews are the best approach for the market research.

Personal interviews are the most expensive way to do research but, as Heather explains, it is the best way to gather information for products that are very technical or complex. RPC's Industrial Product Specialist, Mark DeLong, does numerous personal interviews for his industrial market research projects. Using a prototype of the product or designs of the product, Mark gathers information from various key informants.

For food products, focus group taste tests are a good source of information.

"It is important to get a group made up of the type of person that we want to sell to and ask them how they feel about the product or how they feel about the packaging; what it does for them and what it doesn't do for them," explained Heather. "We usually sample a variety of different products at a taste test and sometimes we compare our client's products to competitive products to get feedback on which products the group prefers and for what reasons: colour, packaging, taste or any number of things.

"From this, we learn which features are important to our target consumer. With this information, we can refer the client to our Food, Fisheries and Aquaculture department for appropriate changes to the product. We can take that new product back to a focus group and make sure that we have not done anything to alter the positive impressions of this product. It's a step by step improvement; ideally you go in and everyone loves your product and then you can do a

(continued on page four)

Figure 4.25. RPC's internal newsletter *"Inside Out."* This issue was published in July 1993.

The need for RPC to provide its Technical Information Services eventually declined, and the entire Information Technology Department was spun-off to the private sector in 1999/2000 [210].

INAUGURAL ISSUE

RPC is pleased to introduce the first issue of our quarterly newsletter to New Brunswick industry. RPC Alert is a vehicle for communicating technological developments and outlining RPC projects and expertise. We welcome requests for inclusion on our mailing list.

RIBBON CANDY FROM COMPUTERS

RPC has designed and built a machine for forming the ribbon shape in candy. The machine replaces a hand-cranked model with specially crafted brass interlocking sections to form the ribbon.

The new machine, from a design created on the CAD (Computer Aided Design) system, is driven by a variable speed DC motor with squeeze rollers to control the candy thickness. The interlocking sections were machined from solid brass using a computer numerical controlled (CNC) milling machine. Enquiries can be directed to P. Waterhouse, Section Head, Mechanical Technology.

INFORMATION FROM AROUND THE WORLD

"Within the last 20 years, 3/4 of all the information available to the world has been developed. Information doubles every 10 years...each year 72 billion pieces of information are disseminated."

The RPC Information Centre has access to hundreds of computerized bibliographic databases dealing with the sciences, technology, business and the humanities. From any number of access points (subject, author, corporate source, keywords) or combinations thereof, a selected database may be "searched" to retrieve references of interest.

Once a suitable set of references has been obtained, clients may begin the process of acquiring copies of the documents cited, usually through inter-library loan. The process may simply involve a check of our own holdings (10,000 volumes and 250 periodicals), or a search on the Phoenix System to establish a location at the University of New Brunswick, or one of the provincial government libraries. If, however, no local facility holds the document, steps are taken to identify an out-of-city location. This may be the Regional Library system; the Canadian Institute of Scientific and Technical Information (CISTI), National Research Council; the Library of Congress, Washington, D.C.; or the British

Library, Lending Division, London, U.K. - to name a few. A request for inter-library loan is then instituted, and depending on the distances involved, the material is received 2-6 weeks later.

These services are available by contacting Valerie Owen, Information Officer.

CONSERVING ENERGY IN THE WORK PLACE – "NO CHARGE" SERVICE TO BUSINESS AND INDUSTRY

RPC has contracted for the delivery of the Canada Energy Audit Program (CEAP) in the Province of New Brunswick. The CEAP, funded by the Department of Energy, Mines and Resources assists industrial and commercial firms and institutional organizations to conserve energy and reduce operating costs. RPC will perform energy studies, identify energy saving opportunities and make recommendations for energy management conservation measures.

RPC operates three teams of energy audit engineers and technicians with three vehicles equipped with computers and instruments for on-site energy audits. The service is offered free of charge to energy consumers. For further information, please contact Rick Kowalski, CEAP Coordinator/Manager.

IRAP-L: HELPING INDUSTRY SOLVE TECHNICAL AND PRODUCTION PROBLEMS

Since 1981, nearly $800,000 has been invested in solving more than 165 engineering and technical problems encountered by New Brunswick's small and medium sized manufacturers.

The source of this funding is a special element of the National Research Council's Industrial Research Assistance Program (IRAP) known as IRAP-L which assists companies in solving short-term engineering and technical problems, improving production operations, and expanding or enhancing their effective technology base through the use of specialists and facilities not available within the company. Any New Brunswick manufacturer with less than 200 employees is eligible to receive IRAP-L assistance for 75 percent of a project's cost, providing the total project cost is no more than $6,000. Projects typically range from the design of new or improved equipment or products, to the testing of completed prototypes.

Figure 4.26. RPC's quarterly publication "*Alert*" launched in December 1984, with technology news, summaries, and updates. It was renamed "*Visions*" and made triannual in 1994.

In 1990, RPC adopted a modernized version of its signature logo, combined with the descriptor: "Technology and innovation specialists" (Figure 4.27). The lettering adopted is called Microgramma and its "sleek futuristic style" was intended to identify a corporation committed to technological innovation [211,212].

Fish diagnostic kit a commercial reality

A microbiologist tests an Atlantic salmon for disease using antibody-linked latex bead technology.

Negotiations have been concluded with Valox Ltd. of Fredericton, New Brunswick for commercializing the innovative Agglutitest technology which has been developed by RPC and the University of New Brunswick (U.N.B.) for the early detection of serious diseases, such as Bacterial Kidney Disease and Furunculosis, in farmed salmonids. Utilizing antibody linked latex beads and a microcapillary procedure, the Agglutitest kits provide salmon farmers with a sensitive, inexpensive and user-friendly kit format for screening their stocks of fish. Under license from RPC, Valox Ltd. will be manufacturing and distributing the Agglutitest kits to customers in North America, South America, Europe and Asia.

Currently, RPC and UNB scientists are assisting Valox Ltd. in the technology validation process which involves demonstrating the technology to representative customers in the Maritimes and British Columbia. Validation provides an opportunity for input from customers in refining the ergonomics of the product so that they will be packaged appropriately for the marketplace. The one year validation process, which will involve testing for the presence of diseases in both the freshwater and marine phases of Atlantic salmon, as well as in Pacific salmon and various species of trout, will be completed in September, 1991. At this time, Valox Ltd. will be manufacturing and marketing the products through its network of distributors.

In this issue:

Ideas - the building blocks of successful research

An "Ideas Group" has been formed at RPC to recommend innovative research ventures. In 1990, six research projects have been set in motion based on the recommendations of the Ideas Group.

Some of the research projects underway at RPC are: ultrasonic evaluation of concrete deterioration; electrochemical synthesis of hydroxynicotine; and the development of Optoelectronic Cross-Coupled Addressable Matrix (OCCAM) technology

The Ideas Group and related RPC staff have found the organization's solid background in organic chemistry to be a strong asset for the development of electrochemical synthesis for hydroxy-nicotine. RPC is conducting research to determine the viability of an electro-chemical synthesis of 2-hydroxynicotinic acid. If the research indicates feasibility, RPC will be looking to fund a joint study with an outside partner. This would, in turn, lead to the active participation and funding of a third partner for the development of a commercial scale production process.

OCCAM technology, the focus of another research project underway at RPC, is primarily applied by military and aerospace, non-destructive examination, robotic, process sensing and mining markets. The concept of OCCAM springs from a combustion of various fiber-optic sensor technologies including a patented proprietary fiber-optic level sensing methodology. OCCAM permits sensing systems for parameters including pressure, stress, temperature and damage to be built into a structure such as a wing, building, robotic end effector or a space station. Initial work with OCCAM technology involves investigation into various sensor types and identification of applications for this technology. *(Continued on page two)*

(Mailing Label)
(Étiquette d'adresse)

If you are able to read this, you are not on our mailing list. To receive Alert regularly, see the opposite cover.

Si vous lisez ceci, c'est que vous ne faites pas partie de notre liste de distribution. Pour recevoir Alerte régulièrement, voyez notre page 22 de couverture.

Figure 4.27. In 1990 RPC updated its company logo, as seen on this Winter 1990 cover of RPC's "*Alert*" newsletter (upper left corner).

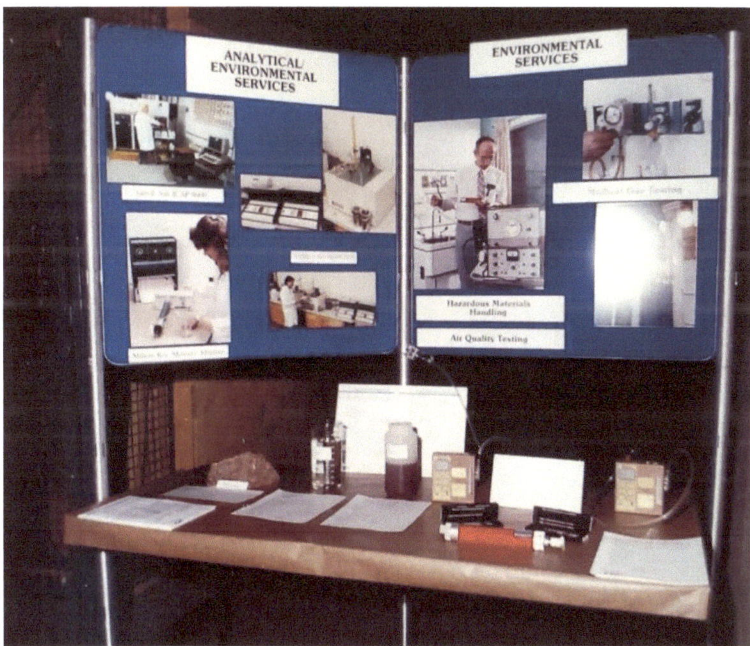

Figure 4.28. **RPC open house on February 20, 1990. Upper Photo: Premier Frank McKenna (left) with Dr. Boorman (centre) and Dr. Grotterod (right). Lower Photo: One of the many displays. RPC photos.**

Figure 4.29. Air Quality: RPC's Thelma Green testing for asbestos. RPC photos, *circa* 1990.

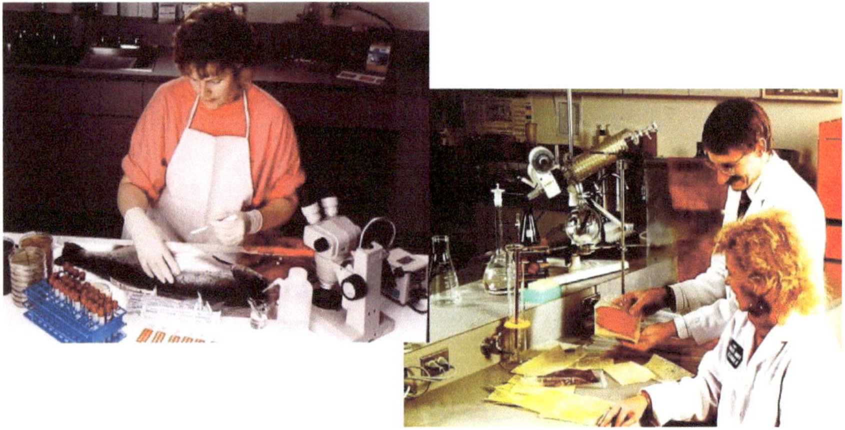

Figure 4.30. Fish Health with RPC's Lynn Hutchin (Left image). Food Product Development with RPC's Carol Crouse (seated) and Germain Landry (standing) in the right-hand image. RPC photos, *circa* 1990.

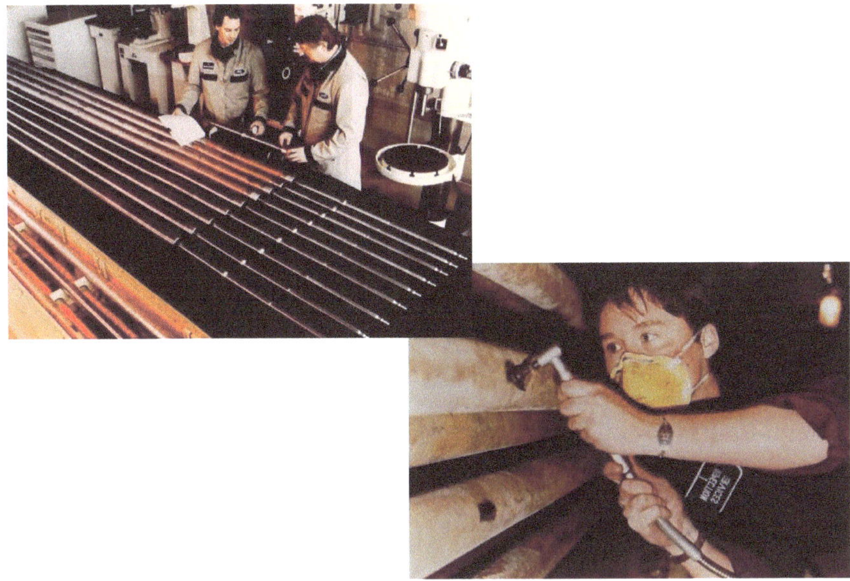

Figure 4.31. Control Rod Engineering with RPC's John Aikens and Greg Brown (Left image). Physical Metallurgy with RPC's Maribeth MacNutt (Right-hand image). RPC photos, *circa* 1990.

In 1992, Dr. Boorman retired from RPC. In his stead, Dr. Peter Lewell[39] was promoted to become RPC's third Permanent Head [158] (Figure 4.32). He served in this capacity, as Executive Director of RPC, until September 2004 [213] (See Appendix 8.3).

In 1992, RPC celebrated its 30th Anniversary. To mark the occasion, RPC clients, Board of Directors, government officials, and media were invited to "A Showcase of Innovation," on June 4, 1992.

---

[39] Dr. Lewell had been with RPC since September 1969, and had served as Deputy Director since 1986. He was made Acting Executive Director in July 1991 (in Dr. Boorman's temporary absence) and officially appointed Executive Director and as a Board Member in 1992.

Figure 4.32. RPC's third permanent Executive Director, Dr. Peter Lewell took office in 1992 [158].

The anniversary showcase included innovation awards to clients[40], workshops, and a buffet lunch of client-produced products. In an editorial titled "RPC Has Key Role to Play," the *Fredericton Daily Gleaner* praised RPC's work with small companies and entrepreneurs [214], concluding that:

"The work of RPC in supporting those with a creative and entrepreneurial spirit is necessary… [and] Their effort in attracting new businesses and supporting existing ones is certainly a most significant move in the economic life of the province right now."

---

[40] These were presented to RPC clients that have shown leadership within the previous 30 years and had created commercial successes.

As RPC advanced beyond it's 30th anniversary, it progressed in many areas. The following descriptions provide only a very few examples:

Centre for Nuclear Energy Research (CNER). RPC had experienced some successes in creating joint ventures[41] with UNB (see Appendix 8.5) and in 1992 they came together to create another one. CNER was established in 1992, in partnership with UNB and NB Power. Its purpose was to *"bring together research scientists from UNB, RPC and other parties to conduct individual and collaborative research and development (R&D) in specific areas related to the safety and operation of nuclear generating plants. By special arrangement, researchers in the centre will have access to facilities at UNB, RPC and NB Power's Point Lepreau Generating Station"* [215]. AECL Research, a subsidiary of AECL, became the centre's first collaborative research partner and announced that they were moving six of their scientists and engineers to Fredericton to work at CNER [216]. In December 2010, CNER was reorganized as a research institute within UNB, with RPC shifting from being a co-owner/operator to a collaborator [217,218].

Food Science and Tasting Panels. RPC's tasting panels (see also Chapters 2 and 3) continued to form an important component of its work in consumer food product development for regional companies. In a 1993 issue of RPC's internal newsletter *Inside Out*, RPC's Consumer Product Specialist Heather Wilson described the service [219]. Heather explained that she dealt with two different types of clients: clients that have a real product and clients whose product is at the idea stage that hire RPC to determine if the idea can be developed into a product. Depending on which category the client fits, one of two different approaches is taken:

"For someone with an actual product, our job is to validate that all of the features on the product meet consumer demands… So, our first job is to identify what consumer group would be interested in buying this product and then seek those people out to ask them questions about whether or not this product would interest them and if the features in the product meet their needs."

If RPC learns that the features do not meet consumer needs, they identify what changes are necessary to create a demand for the product. The consumer group is typically made up of individuals that represent the target customer market for the product:

---

[41] Joint ventures meaning jointly founded and operated in practice, and not necessarily meeting the definition of legal joint-ventures.

"Since the second [type of client] does not have a product at this stage, we take their concept and develop it. The client often needs us to define what features the product should have, what price it should be, or where it should be sold and we determine those answers through market research," Heather added, and here too, focus group taste tests are used, sometimes using prototype products. *"It is important to get a group made up of the type of person that we want to sell to and ask them how they feel about the product or how they feel about the packaging; what it does for them and what it doesn't do for them… We usually sample a variety of different products at a taste test and sometimes we compare our client's products to competitive products to get feedback on which products the group prefers and for what reasons: colour, packaging, taste or any number of things. From this, we learn which features are important to our target consumer. With this information, we can… [make] appropriate changes to the product. We can take that new product back to a focus group and make sure that we have not done anything to alter the positive impressions of this product. It's a step by step improvement; ideally you go in and everyone loves your product and then you can do a test launch for the product…"*

Bioremediation Services. In 1992, RPC launched bioremediation bench-testing and feasibility services to environmental consulting firms such as Geobac, Maritime Groundwater, and Jacques Whitford [220]. RPC had already been offering analytical-chemical services to this sector for several years, and the 1992 addition added microbiological expertise and testing. Bioremediation refers to the microbiological decontamination of soils, in which bacteria are used to break down the contaminants. An early such program was conducted for the N.B. Department of the Environment, in which RPC examined *"the use of white rot fungi to degrade high molecular weight polyaromatic hydrocarbons (PAHs) which are normally difficult to degrade"* [220]. This kind of work was applied to both 'land farming' and '*in situ* bioremediation' of contaminated sites. In the former, the contaminated soils are actually excavated and transported to a treatment site. In the former, wells are drilled in and around the contaminated area, and used to both inject and monitor bacteria, nutrients, oxygen and (usually) water.

In 1995/96, RPC did additional such work (through its Minuvar subsidiary), involving bench-scale testing of a low-temperature thermal-desorption (LTTD) system for the remediation of petroleum-fuel-contaminated soils containing PAHs and pentachlorophenol (PCP). This work, done for Envirosoil Ltd., was in support of an environmental remediation project in Nova Scotia.

Bio-Leaching Processes. Closely related to the bioremediation services just summarized is RPC's entry into the development of *in situ* and 'heap' mineral processing technologies employing bacteria (usually involving bacterial oxidation). Such bio-leaching processes can offer the potential for more effective, and also more environmentally sustainable, industrial metals recovery than conventional alternatives[42]. By 1992/93, this was already recognized as a 'growth area' for RPC, building on its work on bio-leaching processes for copper recovery from mineral deposits [221]. By 1996/97, the client list for bio-leach process development services had expanded beyond the province and beyond the North American continent [222]. In fact, by 1999, RPC had entered into a strategic alliance with Pacific Ore Technologies (Perth, Australia) to conduct laboratory- and pilot-scale studies of bio-leach recovery processes on mineral samples from around the globe [210]. In 2000, the latter involved a 5,000 tonne heap-leach pilot plant at one site, and a 7000 tonne one at another [175]. (See *"Titan Trials BIOHEAP™"* in Chapter 6, and the patents by Hunter *et al.* in Appendix 8.8.)

Air Quality Services. In a 1994 issue of RPC's internal newsletter *Inside Out,* RPC Chemist Thelma Green provided an example of their air quality services [223], pointing out that most air quality problems do not have the "*good manners*" to restrict themselves to the traditional 9 to 5 workday. During the summer of 1994, for example, RPC was contracted to be on-call, 24-hours-a-day, at the Irving Pulp and Paper Mill in Saint John during a boiler change-over:

"We were doing asbestos monitoring, identification, and air sampling when it was warranted. If the workers found asbestos we would do the air sampling to make sure that the area was okay," explained Thelma. *"I was there two years ago during the initial stages of the boiler change-over. From that analysis, they know that asbestos is present in the plant but since its location is not documented, no one knew where they might come across asbestos during the shutdown."*

Thelma described the boiler as being like a gigantic (12-storey) elevator shaft with tubes around the outside. *"It gets extremely hot with the steam going through it so a lot of insulation is needed for protection…"* During [the] removal of old boiler tubes and installation of new boiler tubes, RPC staff collected samples from the boiler when needed and analysed them in their portable laboratory. *"I had all my microscopes there to identify samples. If it was asbestos they would call in the asbestos removal company… and have it removed. Afterwards I would do clearance air samples to make sure that the air was safe so the 1,500 contractors and the Irving employees could return to work in that area."*

The foregoing is just one of many examples of public and private concerns about air quality, whether indoor or outdoor, that occur every year. Sometimes, when RPC is called in, sampling and analysis shows that there is no real health or environmental problem, other times a specific cause or agent

---

[42] Such processes can have lower capital costs and the potential to be economic for low-grade ore bodies, with reduced energy use, water pollution, and greenhouse gas emissions.

is identified, which can be addressed directly. In many cases, public concerns have fuelled eye-catching headlines and increased public concern, especially when schools and hospitals are involved. Early in 1998, for example, a headline announced *"Source of Mystery Odour in Lab at Chalmers Remains Unknown"* [224]. In this case, the microbiology lab of the Dr. Everett Chalmers Hospital was plagued by strange smells that resulted in some staff members being sent home sick and the lab itself being shut-down for a week despite having cleaned the drain and ventilation systems, and having replaced duct work and glass pipes in the lab [224,225]. Sampling campaigns conducted by RPC, however, were able to trace the problem to a blockage in a drainage pipe shared by two different labs. By modifying the drainage system so that each lab had its own piping the problem was solved and the laboratory re-opened [225].

RPC and SMEs. A principal focus during *The Commercial Years*, and one that was strongly championed by Dr. Lewell, was in maximizing RPC's engagement with the province's entrepreneurs and its small- and medium-sized enterprises (SMEs)[43]. This engagement had been steadily growing, and Dr. Lewell was determined to keep it growing.

The 1984/85 Annual Report noted that 80% of RPC's 200 clients that year had been firms employing less than 200 persons [149]. Beginning with the 1985/86 fiscal year, RPC began reporting annual revenues from SMEs and tracking the proportion of revenue deriving from SMEs versus other kinds of industrial clients. In 1984/85, RPC served 160 SME clients. By 1989 through 1991 it had risen to the 260 range, and from 1996/97 onwards, it was consistently over 300.

In 1997/98, RPC began assisting (mostly SME) clients with the development of new software products [226].

Extra-Provincial Contract Work. Another feature of *The Commercial Years*, was in diversifying RPC's sources of contract revenue. Taking on contract projects and programs from extra-provincial sources was not new for RPC. As noted in earlier chapters, RPC had been providing training programs in Nova Scotia as early as 1963, followed by counterparts in PEI, and it had also begun partnering with the National Research Council to develop and deliver technical information and field engineering services to industry at about the same time. What was new, was the growth in extra-provincial work. Between 1991/92 and 1996/97, for example, RPC performed work for industrial clients outside of N.B., averaging approximately $1.75 million per year [222] out of total revenues averaging

---

[43] There are various definitions for the sizes of small-, medium-, and larger-scale organizations. RPC has defined SMEs as enterprises having fewer than 200 persons.

approximately $8.28 million per year, meaning that extra-provincial contracts represented about 21% of total revenue. The nature of this work varied considerably from year to year. For example, Dr. Lewell wrote, in 1996/97, that, *"A 'biotech' theme [was] associated with much of the work [RPC performed] for clients outside New Brunswick, including development of vaccines, biopesticides and bioleach technology"* [222]. In 1997/98, *"Clients in power generation, mineral development, specialty chemical and heavy equipment manufacture accounted for much of [RPC's] industrial work outside New Brunswick"* [226].

Among the reasons that regional RTOs typically accept, or even pursue extra-provincial work are that it:

o **Enables critical-mass:** in regions or sectors that do not yet have a high concentration of industry, then the cost of the critical mass (in terms of people, technologies, instruments, and infrastructure) needed to support local industry generally exceeds the revenue that can be generated locally. Doing extra-provincial work can provide sufficient revenue to close the gap,

o **Maintains 'leading edge':** working extra-provincially, and especially internationally, helps ensure that the RTOs know-how and service capabilities are leading-edge and world-class, enabling the RTO to deliver demonstrably leading-edge, world-class support to local industries,

o **Enables technology transfer:** working extra-provincially exposes the RTO to the best practises of companies around the world. This knowledge can then be brought back home where it can be adapted and adopted by local companies,

o **Creates access to outside capital:** sometimes when a home-grown technology is ready for scale-up from pilot- to demonstration-scale, the availability of capital means that it is necessary to partner with other organizations, in other jurisdictions, and conduct the demonstration-scale work there. The benefit is that once the technology has been successfully demonstrated at scale, it can be brought home for commercialization,

o **Brings money into the local region:** when RTOs work extra-provincially, they usually do so at their highest charge-rates, enabling both revenue and net income to flow back to their own region. For RTOs with a low proportion of core operating funding from government, this is sometimes the only way for the RTO to pursue their Mandate while remaining a going concern.

By the end of *The Commercial Years*, in 2004, work performed for extra-provincial clients had risen to slightly more than $2 million per year, representing approximately 26% of total revenue [213].

Incutech. In 1988, RPC and UNB came together to create a joint venture focused on entrepreneurs and SMEs (Figure 4.33 and 4.34; see Appendix 8.5). The new joint venture was called Incutech Brunswick Inc., and was sponsored by the Canada/New Brunswick Subsidiary Agreement on Industrial Innovation and Technology Development. The aim of Incutech was to pilot test/demonstrate a new-venture-incubation centre, and support *"entrepreneurs who are seeking to establish a technology- intensive manufacturing business in all regions of the province"* [150,227]. This was New Brunswick's first business incubator/accelerator facility (Figure 4.34).

Figure 4.33. Newspaper ads explaining and promoting Incutech, an RPC-UNB joint venture. From the *Moncton Times Transcript,* March 18, 1989, p.35 (left) and June 5, 1989 (right).

Figure 4.34. One of the new start-up firms residing in Incutech circa 1989 [228].

Incutech's Manager, Don Doucett, explained that: "*About 80 per cent of new firms fail in the first two years. We are trying to improve the survival rate*" [229]. One of Incutech's tools for this was establishing a support community involving service companies such as "*a business forms company, a graphics designer, and a computer training firm*" [229].

Incutech was successful. By 1991, its client tenants included: Marcel Enterprises Ltd., (power-unit machines for tree harvesting), Mynic Inc. (electronic data collectors), E.M. Technologies (electromagnetic testing laboratories), Energy Recover Corp. (engineering services related to microwave technologies), Xylaur Enterprises Ltd. (equipment for companies developing microchips), and Atlantic Systems Group (training and custom software design), among others [230]. In addition to private-sector-generated start-ups, two RPC spin-offs were housed via Incutech in 1997/98 [226], and Incutech itself spun-out EM Technologies and Atlantic On-Line Systems Ltd. in 1995 [231], and Miratech Inc. in 1997/98 [226].

The provincial government, however, had other ideas and ceased funding Incutech in 1992, threatening its existence [232,233]. Ultimately, Incutech's functions were essentially absorbed into UNB, and continued to operate as the university's entrepreneurial development centre, which was rebranded as 'Enterprise UNB' in 2001 [234].

The strong growth in SME clients throughout the 1980s created a solid, sustainable base for RPC's total revenues (see Figure 4.35). By 1993, Dr. Lewell viewed RPC's work for SMEs as being [221]:

"[well] diffused throughout all regions of New Brunswick and through most sectors of the economy… [drawing] on specialized technical expertise in all our departments… [and] continually nurtured by the focused marketing efforts of our SME business development group."

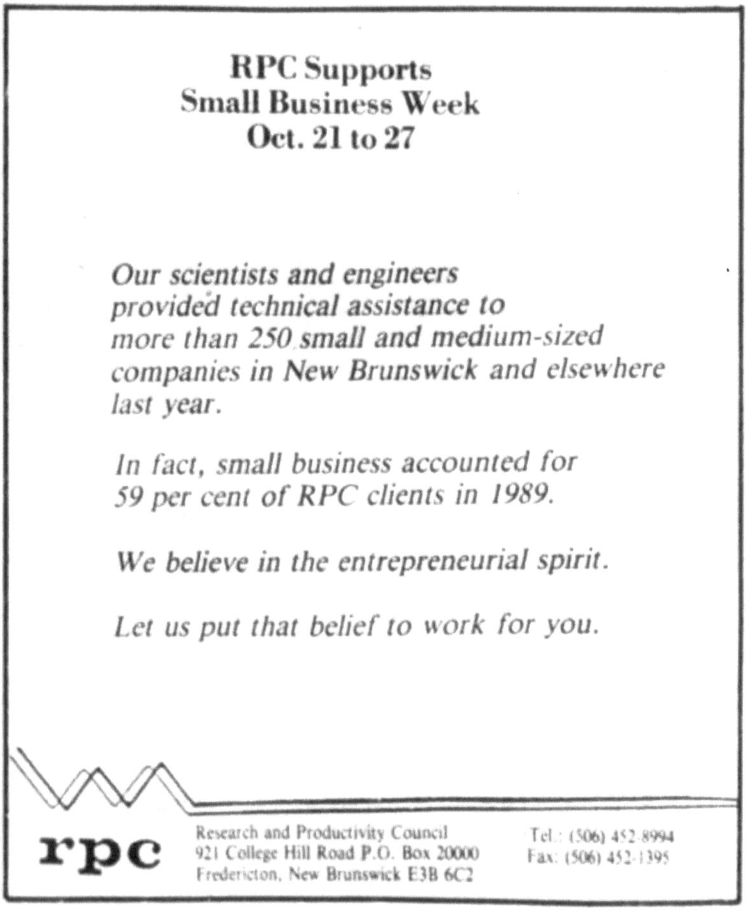

**RPC Supports
Small Business Week
Oct. 21 to 27**

*Our scientists and engineers
provided technical assistance to
more than 250 small and medium-sized
companies in New Brunswick and elsewhere
last year.*

*In fact, small business accounted for
59 per cent of RPC clients in 1989.*

We believe in the entrepreneurial spirit.

Let us put that belief to work for you.

rpc

Research and Productivity Council
921 College Hill Road P.O. Box 20000
Fredericton, New Brunswick E3B 6C2

Tel.: (506) 452-8994
Fax: (506) 452-1395

Figure 4.35. Newspaper ad targeting small businesses. From the New Brunswick Telegraph Journal, October 27, 1990 [236].

One of the reasons for RPC's success with its SME focus was that it built on the many years' work that had been invested in building quality programs for all RPC activities. According to RPC's Hugh Drummond, *"We started getting involved with QA [(Quality Assurance)] in 1975 when the nuclear plant at Point Lepreau was under construction"* [235]. This involved such aspects as purchasing, qualification of suppliers, incoming inspection of materials, final inspection of materials, verification that all necessary procedures are carried out throughout the manufacturing process, and maintaining records so that an individual can tell by looking at the product or looking at the documentation that travels with the product that it has been inspected [235].

RPC's QA program was approved by major clients including SPAR Aerospace, AECL, Ontario Hydro, and NB Power, and individual RPC programs were certified by industry and independent accreditation organizations including the Canadian Association of Environmental Analytical Laboratories (CAEAL), the Standards Council of Canada (SCC), and the Canadian Standards Association (CSA) [235]. Over the years, RPC's programs evolved through newer approaches, such as Total Quality Management (TQM), and international accreditations, such as the International Organization for Standardization (ISO).

A combination of RPC's specialized technical knowledge and capabilities, its linkages with major industry, and its quality programs, enabled the organization to embark on an initiative to 'open doors' for New Brunswick SMEs in international markets.

For example, in 1992/93, RPC qualified as a supplier for the five-billion-dollar CANDU nuclear program in Korea, with the intent of landing contract projects that could then be largely subcontracted to local SMEs. In this way RPC would be able to help them access work they would be unable to pursue independently. Almost immediately, RPC won such a contract, in this case to supply nuclear plant components and engaged a group of New Brunswick SMEs to manufacture the components under RPC's quality assurance umbrella [221,237]. Similarly, RPC won a contract to supply services to the U.S. Spray Drift Task Force, which was led by the U.S. Government, and subcontracted a significant portion of the work to local companies [221].

By 1993/94, RPC was regularly soliciting client feedback upon the completion of projects – a best practice in quality management. In 1993/94, overall client satisfaction with RPC's performance was 90% [237]. These kinds of activities were sufficiently successful that RPC evolved into a leadership position in quality management itself. In 1999/2000 RPC began integrating its various quality accreditations, introduced new internal quality assurance controls, and evolved it all into a single, comprehensive Quality Management System [210].

When the new system was in place and operational, RPC was ready to seek ISO 9001:2000 registration, which was obtained in May 2001 [175,238] (Figure 4.36). As reported by RPC's Chair in 2007/08:

"[RPC's] extensive scope of accreditations provides value to New Brunswick industry which in most cases would otherwise have to obtain accredited services from outside the province" [239].

Figure 4.36. RPC employees gathered to celebrate ISO 9001:2000 registration in May 2001. The official ISO Certificate of Registration is being presented to Dr. Lewell (right) by Mr. Ken Sadler of QMI (left) [238]. RPC photo.

The continuing growth of RPC's physical facilities and equipment is reflected in the million-dollar increase in fixed/capital assets during this era (see Table 4.1).

Summary

In many ways, these were *The Commercial Years*, as RPC positioned itself ever more closely with industrial clients and their needs, and demonstrated that it could deliver practical, quality results within reasonable time periods. Some of this was in response to pressure (and the risk of funding cuts) from RPC's owner. In 1992, for example, the then-Executive-Director Dr. Lewell noted in a media interview that:

"With provincial revenues being down and with their budget problems, support for RPC is down. There is now increased pressure on us to become self-sufficient" [240].

As a result of its commercial focus, the client roster continued to build year by year (Figure 4.37). From approximately 300 in 1983, it had risen to 400 by 1985/86 [160], to 459 by 1990 [204], and it continued to rise. Between 2000/01 and 2003/04, the annual number of clients served was averaging approximately 850 [3,175,213,241].

Table 4.1. Illustration of some of the changes experienced by RPC during *The Commercial Years, 1983 – 2004*. See also Tables 2.2, 3.2, and 5.1.

	1983	2004
Total Revenue	$4,774 k	$8,746 k
Provincial Grant	$600 k	$741 k
Earned Operating	$4,174 k	$8,005 k
Fixed/Capital Assets (mostly buildings, lab equipment and fixtures)	$1,509 k	$2,528 k
Employees	103	94
Clients Served	~300	830
Reference	[145]	[213]

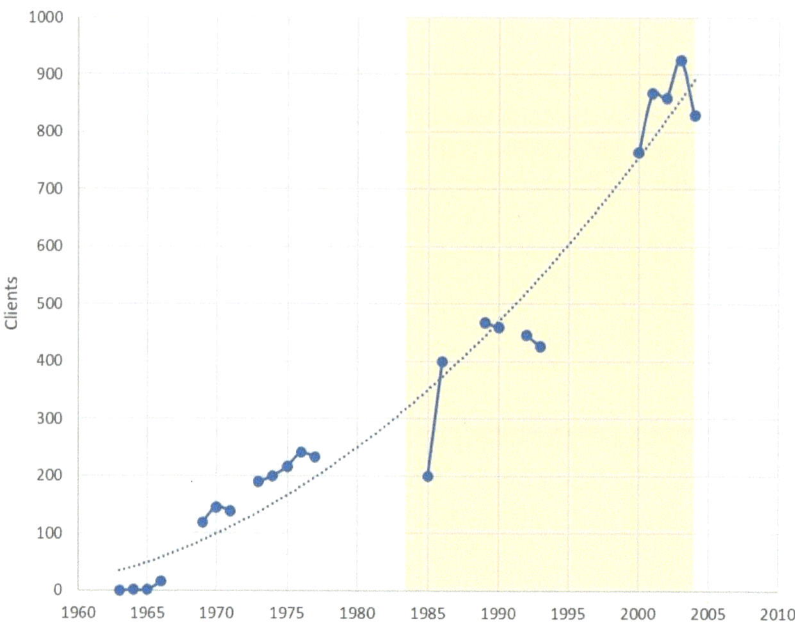

Figure 4.37. RPC annual client numbers to 2004 (fiscal-year-end basis). *The Commercial Years* period is shaded.

During this era 'earned operating income' (defined as total revenue less the provincial grant[44]) averaged 91%. In other words, a dominating focus on clients and contracts had made RPC almost completely market-driven in its outlook. In the next era, *The Entrepreneurial Years*, this would become the salvation of RPC.

As Figure 4.38 shows, revenue growth during *The Commercial Years* was much slower than in the previous era, averaging only 3% per year. As the graph shows, the annual revenues had also become much more volatile, making planning much more difficult.

As a contract services organization, RPC's employee numbers tend to track revenue fairly well, so annual employee numbers experienced the same two trends in this era. That is, slow growth (very slow in this case, averaging only 0.3%) and high volatility, as shown in Figure 4.39. In terms of absolute numbers, the number of employees was actually down slightly by 2003/04, to 94, compared with 103 at the end of 1982/83.

---

[44] This is somewhat distorted by the substantial ACOA grant revenues received between 1988/89 and 1992/93, but not enough to change the average significantly.

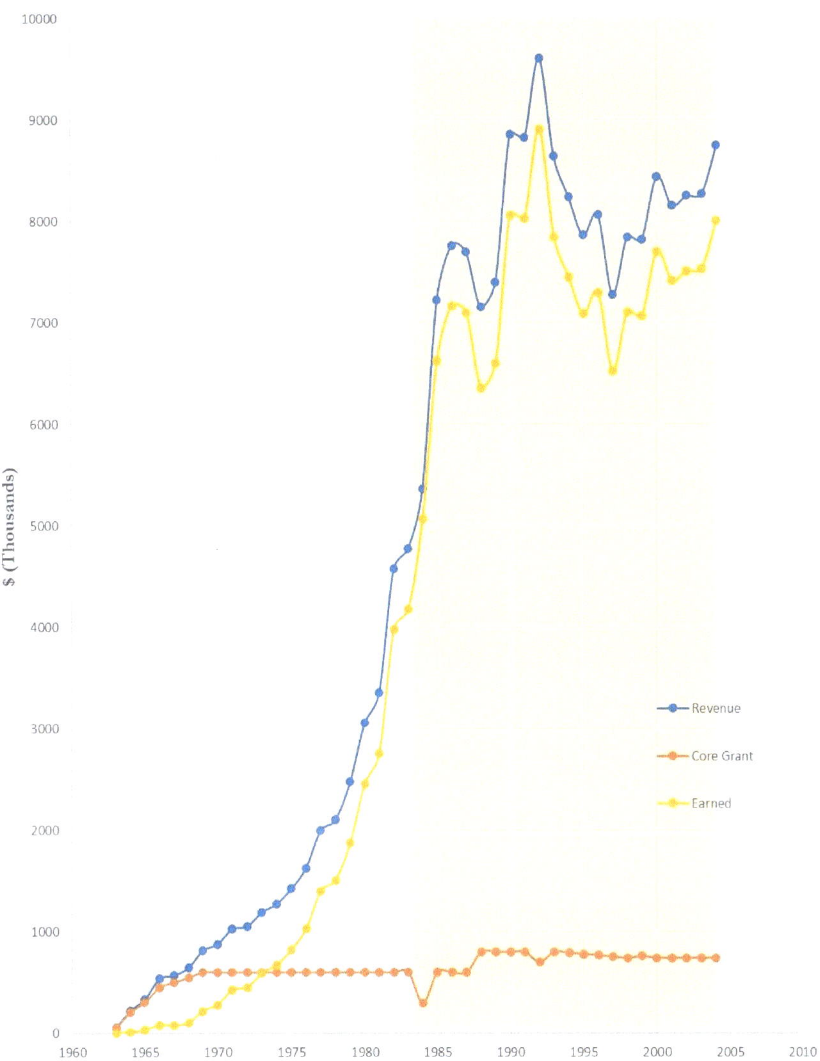

Figure 4.38. RPC revenues to 2004. *The Commercial Years* period is shaded.

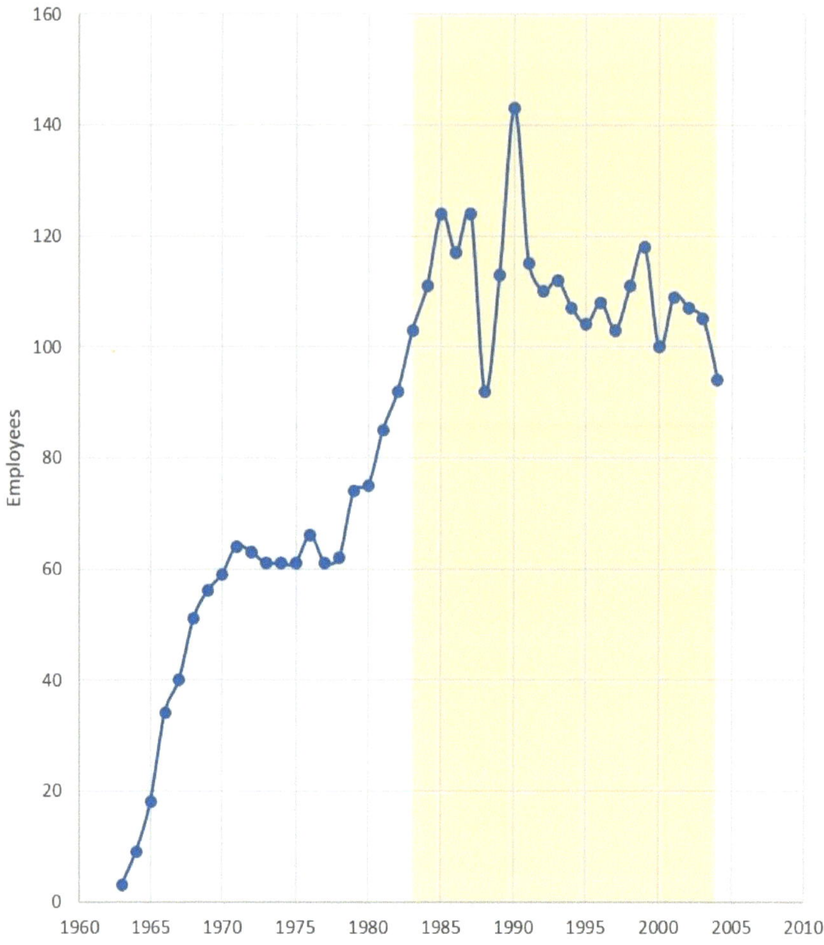

Figure 4.39. RPC employee levels to 2004 (fiscal-year-end basis). *The Commercial Years* period is shaded.

5 THE ENTREPRENEURIAL YEARS, 2004 - 2020

In October 2004, RPC hired Eric Cook, P.Eng., MBA, as its fourth Executive Director [242,243] (Figure 5.1, Appendix 8.3). His initial focus was on renewal and positioning for the future. To enable this, a high degree of importance was placed on a new strategic plan that would aim to increase the level of contract work conducted for industry.

Figure 5.1. Mr. Eric Cook, RPC's fourth permanent Executive Director, took office in 2004 [242].

RPC's First Chief Operating Officer. One of the first changes, was the creation of the position of Chief Operating Officer (COO) to oversee RPC operations including facilities, financials, administration, and common support services. In 2004, long-serving RPC employee Stephen Fox was appointed to be the first COO[45] ([244], Figure 5.2). This expansion of the executive team was intended to help with the renewal and growth of RPC, and helped set the stage for *The Entrepreneurial Years.*

Figure 5.2. Stephen Fox, CA, RPC's first Chief Operating Officer. RPC photo.

Other changes were initiated under a new five-year strategy: "Strategic Plan 2005 – 2009," for which the strategic corporate objectives included [245]:

1. Grow our business,
2. Emphasize our innovation mandate,
3. Re-establish RPC's influential position with the Province of New Brunswick for technical and laboratory services policy,
4. Maintain RPC's position as a leading employer,
5. Develop operations to be versatile, dynamic and agile, and
6. Maintain a stable business.

---

[45] At the time of writing, RPC's Steve Holmes was serving as COO.

The significance of growing RPC's business, under Strategic Goal #1, was to ensure that the organization would make a substantial contribution to the New Brunswick economy, to enable re-investment in the organization, its employees, and its service offerings. To meet this objective, RPC would increase its local market penetration and continue to increase the export of its services across Canada and internationally. In 2005/06, 32% of RPC's clients were based outside of New Brunswick [245], spanning 33 countries beyond Canada (Figure 5.3).

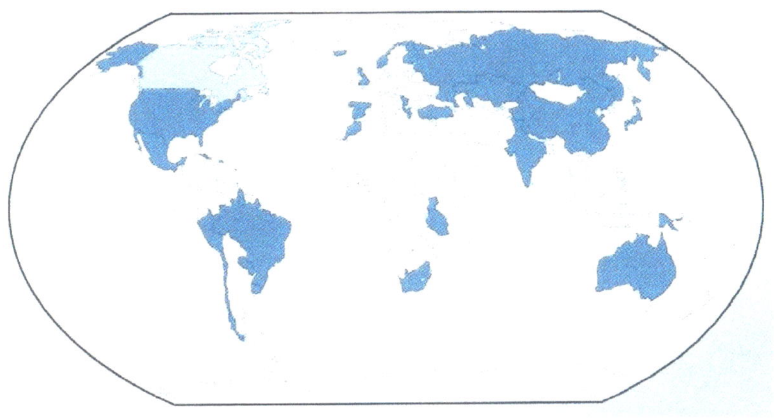

Figure 5.3. Map showing the 33 countries (dark blue shading[46]), beyond Canada, from which RPC clients were based in 2005/06 [245].

Whereas *The Commercial Years* were characterized by a dominating focus on the province's entrepreneurs and SMEs, the new strategic plan focused on industry more broadly, while maintaining (under Strategic Goal #2) support for the technology and technical service needs of New Brunswick's SMEs.

Another key feature of the new strategic plan was the recognition that globalization, shortened product life cycles, and rapidly changing economic and political environments would mean that RPC, and the businesses it intended to serve, would have to be increasingly prepared for change. RPC would aim to be agile: anticipate and respond to change, offer diverse business services, emphasize collaborative efforts where possible, be proactive, and seek to manage the risks inherent in attempting to work on the leading-edge of technology without slipping over the 'bleeding edge.'

---

[46] Albania, Australia, Brazil, Chile, China, Cuba, Dominica, Dominican Republic, Finland, Greece, Guyana, Iceland, India, Japan, Kazakhstan, Mexico, Morocco, South Africa, Norway, Papua New Guinea, Paraguay, Peru, Russia, Serbia, Spain, Tanzania, Trinidad and Tobago, Turkey, Uganda, United Kingdom, United States, Uruguay, and Vietnam.

Strategic Plan 2005 – 2009 laid the foundation for *The Entrepreneurial Years* and although begun with the principles just summarized, it would later become the savior of the organization, as will be seen below.

In the early years of the new strategic plan, three key challenges emerged:

1. the ongoing expansion of national environmental chemistry laboratories into N.B.[47] and elsewhere across Canada,
2. the increasing amount of duplication and redundancy of research and testing services provided by other parts of the provincial government itself, and
3. the decline in core operational funding from the province [246].

The first of these challenges led not only to business competition with RPC, but an exodus of work from the province, as the national and international chains set up storefront operations and shipped most of the samples received to central-Canada labs (using a 'hub-and-spoke' business model). RPC responded to this challenge with continued improvements in customer service, optimizing its value proposition to clients, and promoting the importance of retaining local expertise in environmental chemistry [246]. Not surprisingly, this strategy was well received by local N.B. companies who, for the most part, remained loyal to RPC [239].

The second challenge resulted in the province having over-capacity for testing and analysis in sectors such as environment, the dairy industry, and aquaculture [246]. To address this, RPC responded to this challenge with increased communications – especially to government – to build awareness of RPC's existing capabilities, its value proposition to government, and the critical role it plays in supporting and advancing the N.B. economy. Meanwhile, RPC endeavoured to retain its expertise in the affected areas while diversifying its capabilities and services [246]. Although, as noted in the 2007/08 RPC Annual Report [239], this strategy was not effective at first, it eventually did succeed. In 2017, as a result of the government's Strategic Program Review (SPR, described further below), it transferred to RPC the provincial fish health, dairy, and environment labs, and gave RPC the responsibility for providing laboratory services on behalf of the province.

The third challenge presented the greatest threat to RPC's sustainability. In the words of Dr. Grotterod, RPC's Chair [239]:

"The grant is RPC's lifeblood; cutting it could be devastating to the organization and to the province. As a provincial research organization, we are not empowered to raise capital by traditional means; hence, we rely on the grant and net income for re-

---

[47] A compounding factor was the consolidation of local consulting engineering firms with national engineering firms, which eliminated RPC clients as well as increasing competition.

investment in our human and physical capital. The grant is our only source of government financial support; all other expenses are fully realized by the organization. As New Brunswick's primary resource for applied research and market-led innovation, cutting support for RPC negatively impacts these initiatives that are crucial to a vibrant NB economy, jeopardizes the resource that 600 New Brunswick clients rely on annually and raises doubts about New Brunswick's commitment to innovation."

To address this third challenge, RPC responded with increased communications – especially to government – to build awareness of RPC's existing capabilities, its value proposition to government, and the critical role it plays in supporting and advancing the N.B. economy. This was unsuccessful and, as will be discussed further below, the situation became progressively worse.

Aquaculture and Traceability. A multi-year project to develop DNA-based traceability for farmed salmon began to garner media attention in 2006 [247]. The concept was to use DNA genotyping to determine a DNA 'barcode' for each fish, enabling it to be tracked within a hatchery and throughout all phases of production. RPC's Dr. Rachel Ritchie explained that such technology could *"ensure the safety of food from production to retail outlets and ensure prompt response in case of recalls"* [247].

By 2006, RPC had identified some of the DNA markers and was ready to start developing the software to aid in the tracking. By 2008, RPC had acquired an industry partner in the form of Cooke Aquaculture and, supported by ACOA funding, were proceeding to test the concept of tracking individual fish from 'farm-to-fork' [248].

As a first step, RPC was able to identify unique areas of DNA that are common to the offspring of the broodstock, enabling the tracing of fish back to the parents by batch. While this advance fell short of being able to identify a unique barcode for each fish, it represented a huge technological and commercial step forward. See also "Cooke Aquaculture" in Section 6.4.

In terms of governance, Dr. Grotterod retired from the Board and stepped down as Chairman in 2008, having been RPC's longest serving Chair at 22 years (see Appendix 8.2). In Dr. Grotterod's place, Mr. Ken Reeder[48], was appointed RPC's fourth Chairperson in January 2009 [249] (Figure 5.4).

---

[48] At the time, he had just retired as President and CEO of Neill and Gunter Ltd. (now Stantec).

"RPC thanks Dr. Knut Grotterod for his outstanding contributions to RPC… [He] was appointed RPC Chairperson in 1986 and served in that role until December 2008. During his term, he worked with three Executive Directors and numerous industry leaders who served on the board… Over the years, Knut guided RPC and helped with its growth and success. While there were many accomplishments, perhaps the most significant was his proactive leadership in positioning RPC to excel in the global economy; his foresight in anticipating the opportunities of globalization has helped RPC to prosper.

RPC's 47th Annual Report, June 17, 2009

Figure 5.4. Mr. Ken Reeder, RPC's fourth Chairperson. RPC photo.

A few of RPC's more unusual projects garnered media attention in 2009, including their work testing stun guns and their entry into human DNA testing (Figure 5.5). In February 2009, for example, a bold *Moncton Times Transcript* headline proclaimed: "***N.B. Tests Stun Guns***" [250]. This, in what was possibly one of RPC's more unusual safety-related projects, involved testing conducted-energy weapons, or stun guns, for the N.B. Department of Public Safety to make sure that the units, intended to be used in adult correctional facilities, were properly calibrated, met national standards, and didn't deliver "*a higher amount of electricity than they were supposed to*" [250].

In April 2009, another example was reported, under a similarly bold headline in the *Fredericton Daily Gleaner*: "***RPC First in N.B. to Offer Relationship Analysis***" [251]. What RPC had really begun to offer was an accredited human-DNA-testing service, to establish the presence or absence of parentage. Given RPC's prior work establishing DNA technology for identity and/or traits in fish and animals this was a natural next kind of service for RPC to consider (see also reference [252], and *Aquaculture and Traceability* above, and *Whose Moose is Loose?* and *CSI Deep Woods New Brunswick* below).

Figure 5.5. Headlines announcing some of RPC's more unusual projects in 2009. RPC's Dr. Rachael Ritchie appears in the upper image. From references [251] (upper), and [250] (lower-right).

Supporting New Brunswick Entrepreneurs. The nature of RPC's support for companies has varied somewhat, depending on whether they are individual inventors, entrepreneurs, small- and medium-sized enterprises, large companies, or multinational enterprises. The following is a 2009 sampling of RPC facilitated applied research efforts for New Brunswick inventors and entrepreneurs [249]:

o *Turning Ideas into Reality.* Examples of product development projects during 2008/09 include the application of probiotics to develop antifouling compounds for nets, developing a new environmental process for the recycling of a waste material, development of a bilge water cleaning process, mineral extraction process developments, development of an innovative landscaping material, numerous food processing developments and optimizations, and development of biodegradable bags.

o *Testing New Products.* Consumers, investors, and resellers all want independent validation of new product claims. In New Brunswick, innovators depend on RPC to provide it. During 2008/09, RPC helped N.B. entrepreneurs with testing an innovative exercise machine, a new after-market product for pick-up trucks, an emerging thermal sensor application, and performance verification of a new filter project.

o *Improving Productivity by Extracting Value from Waste.* Waste costs money. When entrepreneurs can develop sales revenue from what was previously considered waste, it has a double impact to the bottom line. RPC facilitated several such successes by researching and developing value-added product uses for lobster roe and a similar project for fish waste. In a related effort, RPC automated a process to remove blood from salmon to assist our client in serving a growing demand for the product from the medical research sector.

o *Developing New Tests.* Creating or preserving value often depends on the ability to detect a parameter. In some cases, a test does not exist and a method to determine the presence or absence of a parameter of significance is required. Innovators count on RPC for this service and during the past year, RPC delivered by developing a quantitative test for a fish pathogen, a DNA-based food allergen test, and a fish virus detection method.

The fish virus detection method just mentioned involved another application of DNA technology, and garnered media attention in 2009 [253] (Figure 5.6). In this case, a polymerase chain reaction (PCR) instrument is used to take a tissue sample and amplify target DNA/RNA sequences that are specific to the virus of interest. This increases the amount of target material in the sample so that, after tagging the selected DNA/RNA sequence with a fluorescent dye, it can be detected. This kind of testing has been applied to fish hatcheries, to help them prevent the spread of such viruses.

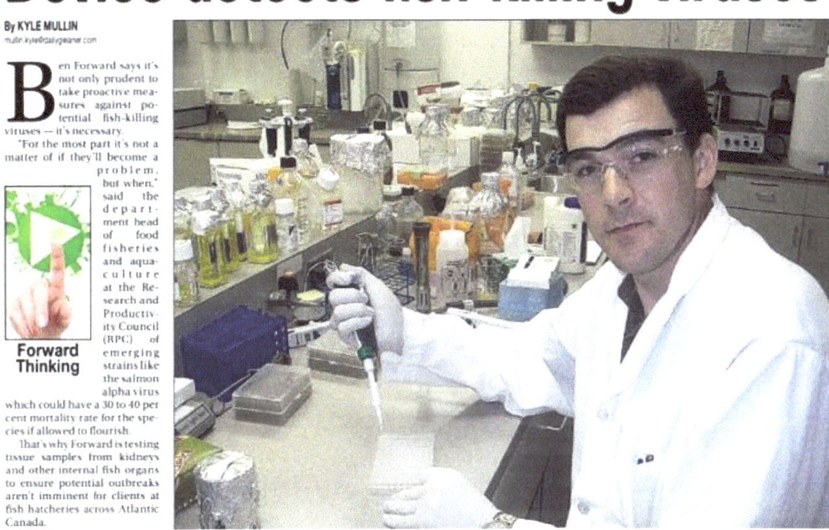

Device detects fish-killing viruses

By KYLE MULLIN
mullin.kyle@dailygleaner.com

Ben Forward says it's not only prudent to take proactive measures against potential fish-killing viruses — it's necessary.

"For the most part it's not a matter of if they'll become a problem, but when," said the department head of food fisheries and aquaculture at the Research and Productivity Council (RPC) of emerging strains like the salmon alpha virus which could have a 30 to 40 per cent mortality rate for the species if allowed to flourish.

That's why Forward is testing tissue samples from kidneys and other internal fish organs to ensure potential outbreaks aren't imminent for clients at fish hatcheries across Atlantic Canada.

Forward Thinking

Figure 5.6. Media article describing RPC's DNA-based fish virus detection service. Shown in the lab is RPC's Dr. Ben Forward. From reference [253] (2009).

Other projects completed for N.B. entrepreneurs have included [254]:

- o Environmental process development for recycling of waste,
- o Process evaluation for high performance water filters,
- o Development and testing of sea-lice treatment processes,
- o Shelf-life studies for the food and beverage industry,
- o Electronics testing for a new remote-control product,
- o Mining process development to extract minerals from smelter slag,
- o Precision flow and temperature measurement devices,
- o Environmental process development for waste reduction,

o   Product process optimization, and
o   Probiotic applications.

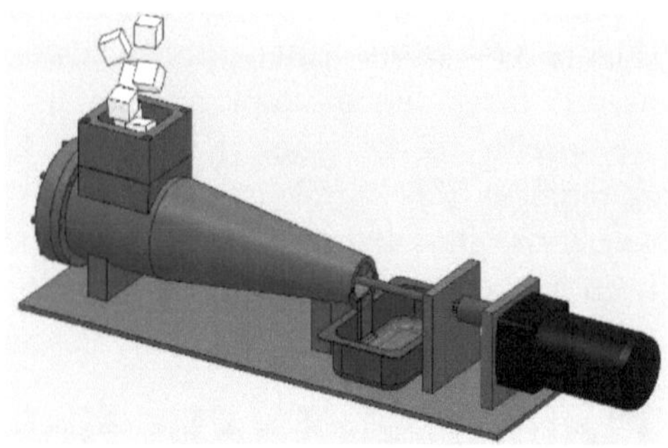

The Decline in Provincial Funding Support. Despite RPC's attempts to persuade the government otherwise, it went ahead with its plan to wind-down its core operational grant-funding support for RPC. For the 2008/09 fiscal year, the province of New Brunswick decreased RPC's core operational grant by 86%, from $740,700 to $100,000 [249]. This was the largest-ever decrease in core operational grant from the province to RPC, which had received an average of approximately $760,000 per year for the preceding 21 years (1987/88 through 2007/08, see Figure 5.17 below). At $100,000, it also represented the lowest level of core operational grant from the province since 1962/63. Although this was a shock to RPC's total revenue, and its ability to invest in its people, tools, and infrastructure, there was worse news yet to come. In the following year, 2009/10, the provincial investment was further reduced to $50,000 [255], and several years later, in 2014/15, it was further reduced to zero [256], where it has remained to the present day [257].

These reductions to the province's core operational grant from the province to RPC seem to have been completely unnoticed by the media.

To fully understand the impact of these changes on RPC, one has to understand why governments invest in RTOs and how RTOs use the money. Governments generally invest in applied R&D, piloting, demonstration, and commercialization because [9,258]:

o   they are direct users of the information and know-how developed,
o   to ensure that strategic economic sectors have a reasonable breadth and depth of R&D that would not otherwise be conducted (because industry either cannot or will not support it on their own), and
o   to ensure that some technologies are developed beyond applied R&D and through such product development steps as prototype development and testing, scale-up engineering, and pilot testing[49].

---

[49] These product development steps comprise the 'bottom' of the *"Valley of Death,"* a regime in which most potential innovations fail and funding is notoriously difficult to obtain [9,258].

RTOs, for their part, generally use their government investment to:
- ○ explore and align with frequently changing market needs, and engage with and enable partnerships and industry consortia in established industrial sectors,
- ○ identify and prepare to meet emerging market needs, and assist business/industry - particularly inventors, entrepreneurs, and SMEs - in emerging areas, and
- ○ continuously build and maintain capacity in order to be able to provide leading-edge technological services to companies *when they need it*. This includes things like continuous training and development of people, market intelligence and foresight, development of new methods, techniques, and tools, and building and equipping leading-edge laboratories, pilot plants, and field operations.

To accomplish all this, RTOs need a certain level of unrestricted government support to supplement their industrial fee-for-service contract work. As discussed elsewhere [9]:

"An international benchmarking study of best practices for RTOs was conducted by the World Association of Industrial and Technological Research Organizations (WAITRO). They concluded that RTOs should receive unrestricted government 'base' funding of 25 to 50% of total revenues, with a 'best practice' level of about 35%... The WAITRO study found that when government investment falls too low, RTOs naturally tend to focus on industrial markets and their short-term needs at the expense of medium- to long-term needs. In this scenario, the amount of innovation enabled falls dramatically. Conversely, when government investment becomes too high, RTOs naturally tend to focus on governments' needs rather than industry's needs. In the extreme case of government funding (90–100%), RTOs tend to focus on just 'Blue-sky' discovery research. In these latter scenarios, the amount of innovation enabled falls dramatically."

At the best-practice, or optimum, level of government funding, RTOs are able to balance the needs of governments and communities with the needs of industries, and to balance the short-, medium-, and long-term needs of both.

In RPC's case, the unrestricted government 'base' funding levels had been trending steadily downward, as a percentage of total revenue, since inception in 1963, having decreased from 100% to 74% during *The Building Years*, from there to 13% during *The Growing Years*, from there to 8% during *The Commercial Years*, and from there to zero by 2015. Although the initial decreases came during eras in which RPC was building its client base and diversifying it, they eventually progressed past the point of best-practice and into the danger zone.

The effects on RPC were rapid, as the organization had to focus more on its top- and bottom-line financials. Beyond business development to replace the lost revenue, increased net income would be needed in order to self-fund the ongoing development of people, infrastructure, and services. RPC was about to become even more entrepreneurial, and this was anticipated in a new five-year strategic plan.

Strategic Plan 2010–2015 focused on the following five strategic corporate objectives [254]:

1. Grow our revenue, invest in our business,
2. Balance our innovation activities,
3. Grow our human capital,
4. Revitalize our relationship with our owner,
5. Develop operations to be versatile, dynamic and agile, and
6. Expand RPC awareness and understanding.

The rationale for these objectives, and the changes from the previous strategic plan will be evident from the foregoing discussion. With funding from the province declining rapidly (see Figure 5.17), it was vitally important to find opportunities to grow the business while building the RPC brand with all stakeholders, particularly its owner.

RPC's principal services at this point in time were categorized as: Technical Services (analyses, tests, and investigations), and Applied Research [254]. The Technical Services had predictable costs, minimal financial risk, and the best financial margins. The Applied Research had greater and less predictable costs, more risk, borderline margins, but stronger alignment with RPC's Mandate and Mission. Therefore, balancing RPC's activities in support of N.B. enterprises' growth would be key.

Growing RPC's Capabilities and Services. The first wave of rapid expansions in capabilities and services included [254]:

o Expanded air quality lab, including new equipment,
o Pilot plant improvements, including a new crusher and an X-ray fluorescence (XRF) instrument that allows for rapid elemental analysis,
o The RPC genetics service line was launched, providing accredited genetic services,
o A new service line in food chemistry was launched, built on newly obtained accreditations for nutritional analysis, and
o The acquisition, early in 2011, of Caduceon Environmental

Laboratories' Moncton location and business[50]. This enabled RPC to both re-establish a facility in Moncton and to conduct accredited microbiological analyses on site. (The new location also served as a depot and local point of contact for RPC's other services.)

The expansion of the air quality lab attracted media attention [259] (Figure 5.7), which enabled RPC to publicly showcase its capabilities in air sample testing for everything from moulds, to asbestos, to radon. A feature of RPC's air quality testing had long been its ability to conduct accredited testing of breathing and medical air[51], the former being critically important to firefighters, divers, and paint-shop workers, among others.

THE DAILY GLEANER/STEPHEN MACGILLIVRAY PHOTO

How good is the air?: Thelma Green, manager of air quality services at the Research and Productivity Centre in Fredericton, is shown doing a breathing air analysis using a gas chromatograph on a sample from a fire department.

Centre expands air testing lab

Safety City facility can test samples for mould, asbestos and radon

Thelma Green, an air quality specialist at the centre, said in an interview the new lab took about six months to build and gives technicians more space and more equipment for

Green said that's because Health Canada has toughened its limits for the odourless gas.

"A lot of people are more concerned now so we have been doing a lot more

Figure 5.7. Media article describing RPC's expanded air quality laboratories. Shown in the lab is RPC's Thelma Green. From reference [259] (2010).

Some of these services have had a long history at RPC. The medical gas testing services, for example, date back at least as far as 1976, when RPC began conducting medical gas testing and inspections for the Dr. Everett Chalmers Hospital [260]. (See the medical gas-testing stories in Section 6.3.)

---

[50] Which had specialized in analyzing environmental samples (see also Appendix 8.5).

[51] See also *"Keeping the Health in our Health Care System"* and *"Health Matters"* in Chapter 6.

In the following two years, 2011/12 and 2012/13, RPC further expanded its analytical service offerings. The Standards Council of Canada (SCC) approved scope extensions to include accreditation for polychlorinated biphenyl (PCB) congeners, a new medical gas standard, a variety of food chemistry parameters, dissolved methane and dissolved ethane[52] [261,262] (Figure 5.8). These increased the accredited analyses available through the air quality group, which included mould, asbestos, and breathing air analysis. Other air quality testing services included radon gas and urea formaldehyde foam insulation (UFFI) surveys [262]. With these in hand, RPC was positioned as having the most extensive suite of accredited analyses 'under one roof' in Atlantic Canada [261].

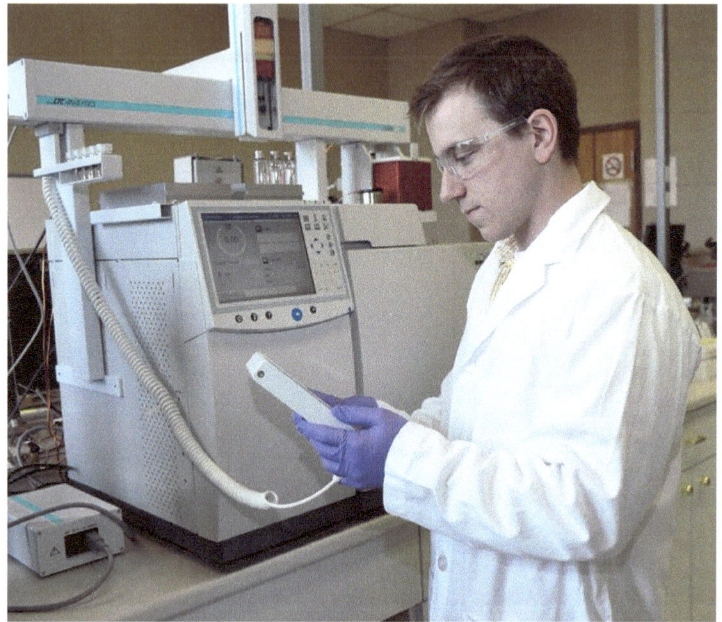

Figure 5.8. RPC's Bryan Bourque, analyzing for Dissolved Methane and Ethane in Water in 2013 [262].

In 2013/14, RPC introduced machine vision services and acquired equipment that allows for proof of concept work to be conducted as the reduced costs and improved functionality of the equipment involved had made more kinds of applications economically viable [263]. Another example was in 3D printing, with the incorporation of a mid-range printer in its

---

[52] PCBs are toxic and resistant to typical environmental degradation. Medical gas piping systems are required to have the system(s) certified through an accredited inspection. Most prepackaged foods sold in retail outlets in Canada require nutritional labeling. Dissolved methane and ethane are potential drinking water contaminants.

product design services[53].

A third example was medical cannabis[54] testing. RPC had begun testing hemp in 1999, under licence from Health Canada. When the Access to Cannabis for Medical Purposes Regulations were updated in 2013, there was immediate demand for analytical testing of cannabis, for which RPC obtained Health Canada licensing in early 2014. RPC also attained accreditation for the parameters required by medical cannabis producers, including inorganic chemistry, organic chemistry, and microbiology. 2016 media reporting placed RPC as the leader in medical cannabis testing [264] which contributed to positively positioning RPC for the Canadian government's legalization of cannabis in 2018, which created cannabis industry demand for additional services, including RPC's Industry 4.0 services, edibles product testing, process testing, genetics and air quality [265,266].

Stormy Times: Arthur Blows the Roof Off. Sometimes safety incidents are beyond a company's control. On what was supposed to have been a quiet summer day Saturday July 5, 2014, Hurricane Arthur blew into Fredericton causing significant damage. RPC was neither immune, nor spared as the extreme wind conditions damaged the roof of Building #3 (Figure 5.9; for reference, see also Figures 2.17 and 4.4). In this case, the winds lifted the entire membrane roof and caused two roof-top HVAC units to flip over.

An employee was in another part of the building, saw the damage and called for help. In a short time, several RPC employees, some with family members, were on-site working to minimize the damage by covering equipment and collecting water that was pouring into the building. RPC's suppliers were very responsive and were on site to take proactive measures for safety, and to prevent further damage until contractors could arrive to shut off the power and secure a temporary repair. The storm was destructive and costly for RPC, and caused some disruption, but no one was injured.

There is no job description or standard operating procedure for what to do when the roof blows off on a Saturday, yet RPC's employees were responsive and proactive, and did the right things to ensure people were safe

---

[53] Which typically starts with computer modelling, followed by printing and manufacturing of prototypes, optimization and testing, then detailed engineering drawings.

[54] In some countries, cannabis is better known as marihuana.

and damage was minimized. In applauding the work of RPC's employees, the 2014/15 Annual Report noted [256] that: *"Every cloud has a silver lining, and in this case, the commitment and care for the organization came shining through."*

Figure 5.9. Roof damage suffered due to Hurricane Arthur in 2014. RPC photo [256].

Between 2007/08 and 2014/15, RPC's annual provincial investment dropped from $740,700 to zero. However, the organization's actions under Strategic Plan 2010–2015 were able to not only replace this revenue but increase it, as total annual revenues over the same period grew from approximately $9.8 million to $10 million (Figure 5.17).

Perhaps most remarkably, although RPC had demonstrated that it could independently survive in the highly-competitive commercial marketplace, it was not prepared to abandon its public-good mandate – the hallmark of a true research and technology organization. This was made explicitly clear in the next strategy, ***Strategic Plan 2015–2020***, which focused on the following five strategic corporate objectives [256]:

1. Revitalization: Position the Organization to Serve its Mandate,
2. Pursue Growth Opportunities,
3. Communications and Business Development,
4. Operational Excellence, and
5. Corporate Social Responsibility.

In order to achieve goal #1, and remain true to its Mandate and Mission, RPC would embark on a revitalization plan to ensure that it was continuing to develop increased capabilities and capacity, in terms of highly-qualified people, leading-edge equipment and services, and appropriate infrastructure. It was anticipated that this would be RPC's most ambitious initiative since the late 1980s [256].

On the governance front, Mr. Reeder retired from the Board and stepped down as Chairman in 2014. In his place, Mr. Lee Corey [55], was appointed RPC's fifth Chairperson in August 2014 [267] (see Appendix 8.2, Figure 5.10). Mr. Corey was well acquainted with RPC, having served on the Board since 2010. Although he would only serve as Chair for two years, Mr. Corey left a lasting impact with his entrepreneurial spirit – that perfectly matched the strategy of the era – and his championing of RPC's capital spending initiative. When Mr. Corey retired from the Board and stepped-down as Chairman in 2016, Dr. Shelley Rinehart [56], was appointed RPC's sixth Chairperson in August 2016 [268] (see Appendix 8.2, Figure 5.11).

As RPC completed its first full year under the new strategy, revenue remained high at over $10 million and the ambitious revitalization initiatives were well underway, all the while maintaining an overall net positive income [269]. RPC remained well positioned as an industry-led organization with substantial government engagement, having realized 83% of revenue from industry contracts and 17% from government contracts. Similarly, although RPC generated a substantial portion of its revenue from extra-provincial contracts, most of its client focus remained dedicated to N.B. clients (52%).

The next substantial increase in RPC's overall analytical capabilities was established as an outcome of the province's Strategic Program Review (SPR). The New Brunswick government had launched SPR on January 13, 2015, aimed at *"finding measures to eliminate the province's deficit"* [270]. Among the resulting decisions were several related to the consolidation of non-medical laboratory services in order to reduce duplication, overlap, and other unnecessary costs [271]. Most of these were implemented in 2016 and formalized in 2017.

On June 19, 2017 the Province proclaimed "An Act Respecting the Consolidation of Certain Laboratories with the New Brunswick Research and Productivity Council" [272], under which RPC became responsible for providing the laboratory services formerly provided by the New Brunswick Analytical Services Laboratory, the Provincial Fish Health Laboratory, and the Provincial Dairy Laboratory (see also Appendix 8.6). Also in 2017, RPC acquired a toxicology service-line from Buchanan Environmental Ltd. ([273,274], see also Appendix 8.6). These acquisitions created several

---

[55] At the time, he was Chairman, Corey Nutrition Company.
[56] At the time, she was Director, MBA Program, UNB Saint John.

opportunities for RPC in 2016/17 [273], including the ability to offer accredited milk testing services, additional fish health services (at its facility in St. George), drinking water testing services on behalf of the provincial government (some through its Moncton facility and others from Fredericton), plus accredited aquatic toxicity testing [273,274].

Figure 5.10. Mr. Lee Corey, RPC's fifth Chairperson. RPC photo.

Meanwhile, RPC's analytical services related to medical cannabis continued to grow, earning RPC a national-leadership position in this area and serving approximately half of the licensed producers in Canada [273].

Figure 5.11. Dr. Shelley Rinehart, RPC's sixth Chairperson. RPC photo.

World-class research is taking place on College Hill Road

LAURIE GUTHRIE
COMMENTARY

I recently had the privilege to tour the Research & Productivity Council (RPC), and was blown away at the magnitude of world-class research taking place on College Hill Road.

With over 1,000 clients in 30 countries from around the world, RPC holds over $15 million in highly sophisticated scientific equipment, and it generates over $10 million annually in revenue. In my

RTOs, and they are affiliated with 350 RTOs in Europe including the largest, Fraunhofer, representing the top 67 research institutes/organizations.

In 1964, Premier Louis Robichaud marked the official sod turning of RPC's home today at 921 College Hill Rd. RPC was created to serve the science, engineering and innovation needs of business and government.

Mr. Cook made an interesting observation about the misuse of the term "innovation." Simply stated, he said innov-

Above is the Research & Productivity Council building on College Hill Road. PHOTO SUBMITTED

of implementation/commercialization. When industry/business pays RPC to conduct innovation research and engineering, they retain intellectual property

proficiency tests annually to maintain their accreditations. Proficiency samples arrive as unknowns from a third party, are tested and reported by RPC and the

Figure 5.12. Media Report Extolling RPC's Virtues in 2016 [275].

Figure 5.13. Illustration of some of the main sectors served by RPC in 2018/19. Clockwise from the top-right: energy, food and beverage, mining, agriculture, aquaculture, dairy, and cannabis. (From the 2018/19 Annual Report [257].)

RPC's First Chief Science Officer. In 2017 and 2018, a "new wave" appeared, in which the federal government and some provinces announced the creation of Chief Scientist or Chief Science Advisor positions across the country (e.g. references [276,277]), accompanied by recommendations for the establishment of such positions in major federal government line departments [278]. Such positions have great potential for RTOs as well, in terms of both internal leadership and external liaison and market intelligence. RPC was among the first RTOs in Canada to create such as position.

In 2018, RPC created the position of Chief Science Officer (CSO) to lead RPC's strategic science direction, coordinate and build consensus internally, and serve as the outward face of science for RPC, building collaborative opportunities, especially with the academic community [279]. Dr. Diane Botelho was appointed as RPC's first CSO ([279], Figure 5.14).

Figure 5.14. Dr. Diane Botelho, RPC's first Chief Science Officer. RPC photo.

The other significant growth area arose from the increasing integration of robotics, machine vision, 3D printing and the Internet of Things (IoT).

Industry 4.0: Helping Manufacturers to Compete. 'Industry 4.0' is the next phase of the industrial revolution. Whereas the first phase of the revolution involved mechanization, water, and steam power (in the late 1700s through to the 1800s), 'Industry 2.0' involved mass production, electrification, and assembly lines, and 'Industry 3.0' involved computerization and automation (Figure 5.15). Industry 4.0 involves digitization. This emerging era is expected to enable 'smart factories,' allowing business to be more agile, offer more personalized products, and make significant productivity and efficiency improvements. The internet, wireless sensors, and software are some of the principal tools of digitization. In 2016/17 [273], RPC made a number of additions in order to facilitate the evolution of Industry 4.0, including an automation lab equipped with robots, machine vision cameras, specialized lighting and conveyors, control systems and sensors. Additional enhancements included incorporating solid modelling and 3D printing into its design processes, developing products that are connected to the internet (i.e., the 'Internet of Things' or IoT), and collecting and monitoring data to improve processes.

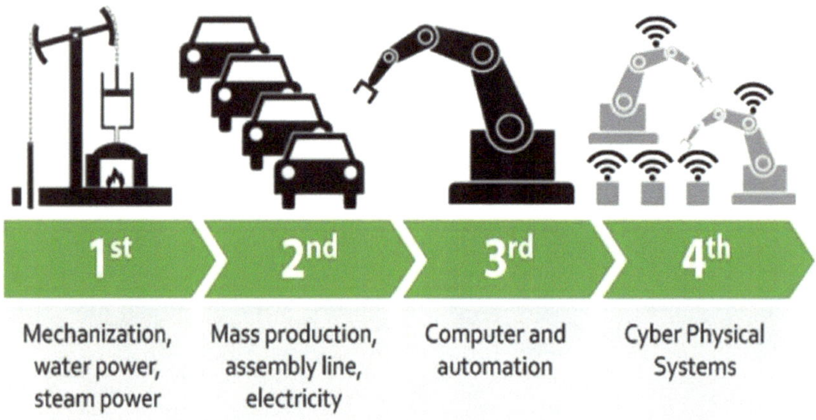

Figure 5.15. Illustration of industrial revolution phases corresponding to Industry 1.0, 2.0, 3.0, and 4.0, respectively. From [273].

Technology advancements in machine learning, machine vision, robotics and other automation technologies have exponentially increased the number of feasible applications. Beyond that, the technology is more functional, more user friendly and more affordable leading to numerous opportunities for SMEs.

In 2018/19, RPC launched an advanced manufacturing technology adoption program with the collaboration and support of ACOA [265]. This program consists of an on-site assessment to identify opportunities for automation, robotics, machine vision, data collection, process automation, data analytics, 3D printing, machine learning and other advanced manufacturing technologies that can help business to improve productivity, quality and overall profitability. As a service provider, RPC can also develop a proof of concept to demonstrate the functionality and performance of the proposed technology, and can assist with a turnkey implementation to integrate the technology into the client's production.

Services for Government. Although RPC's focus, and the majority of its activities (approximately 80% in 2018/19 [280]) are devoted to supporting business and industry, RPC also serves municipal, provincial, and federal governments. As a provincial Crown Corporation, primary among this client sector is the government of New Brunswick, for which RPC's services include routine lab services, such as monitoring and analysis to support regulatory requirements, and crisis response testing, such as that required during spring flooding events. Much of the work done to support the mandates of provincial government departments, however, also benefits business and industry. Examples of this include: fish health analyses in support of the aquaculture industry, milk testing in support of the dairy industry, water testing to support health, tourism and environment, environmental monitoring tests to support construction, mining and other activities, and mould testing supporting indoor air quality.

Summary

In many ways, these were *The Entrepreneurial Years*, as RPC unexpectedly had to deal with the loss of its core operational grant from the province. This has happened to other RTOs in Canada and around the world, from time to time. However, whereas in other cases the affected RTOs have generally been privatized and/or gone out of business, this has not happened in RPC's case. Instead, the provincial government has maintained RPC as a Crown Corporation and RPC, for its part, has resisted the temptation to go completely private and for-profit. Instead, RPC's strategy – in effect - has been as follows:

o   continue to operate as a Crown Corporation, on an overall not-for-profit basis,
o   pursue additional business with above-average margins in order to reinvest the returns in the core capabilities of people, tools, infrastructure, and services, and
o   continue to honour and fulfil RPC's original Mandate and Mission.

In order to accomplish these objectives, RPC has become more entrepreneurial, seeking out new opportunities in response to industry demand (such as in cannabis testing), working closely with government to create opportunities (such as the lab consolidation following the SPR process), and using leading-edge technology to create new business opportunities in N.B. (such as using DNA genotyping for fish traceability, wildlife forensics, and human parentage testing).

As a result of these business development initiatives, RPC was able to keep building its client list almost every single year (Figure 5.16). From 830 in 2004, it rose rapidly to 1,127 by 2018/19 [280].

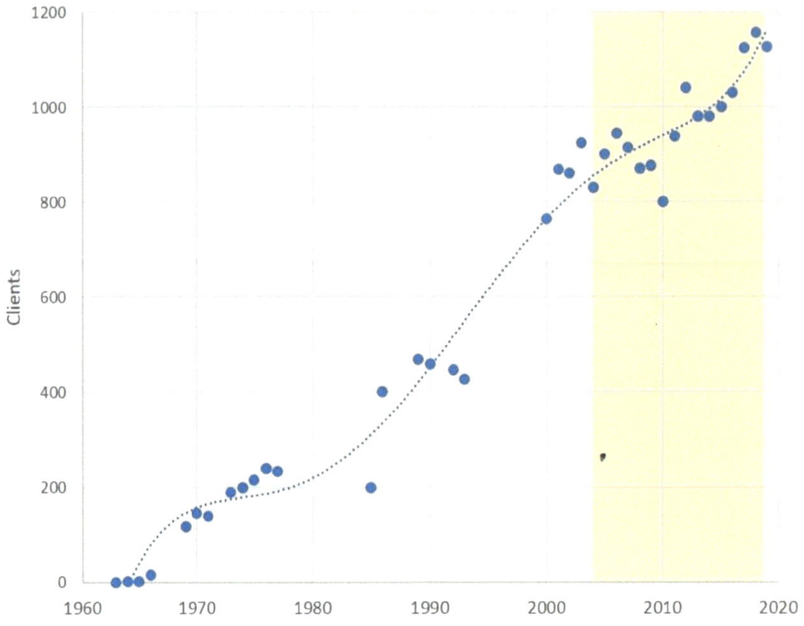

Figure 5.16. RPC annual client numbers to 2019 (fiscal-year-end basis). *The Entrepreneurial Years* **period is shaded.**

RPC's success in developing and launching new services and increasing its client numbers translated into growing revenue over this period. In fact, not only was it able to rapidly replace the lost provincial core operating grant with contract revenues (from both industry and government clients), but total revenue grew at a rate that eclipsed even that of the *The Growing Years* (Figure 5.17), reaching, in 2018/19 a record $16.6 million.

Also, during this period RPC was able to avoid overall staff reductions as the province's core operational funding declined (Figure 5.18). In the most recent three years the staff level dramatically increased so that, compared with the 94 employees that were in place at the end of 2003/04, RPC had an all-time record 160 employees in place at the close of 2018/19 [257].

Table 5.1 summarizes some of the changes experienced by RPC during *The Entrepreneurial Years*, 2004 – 2020.

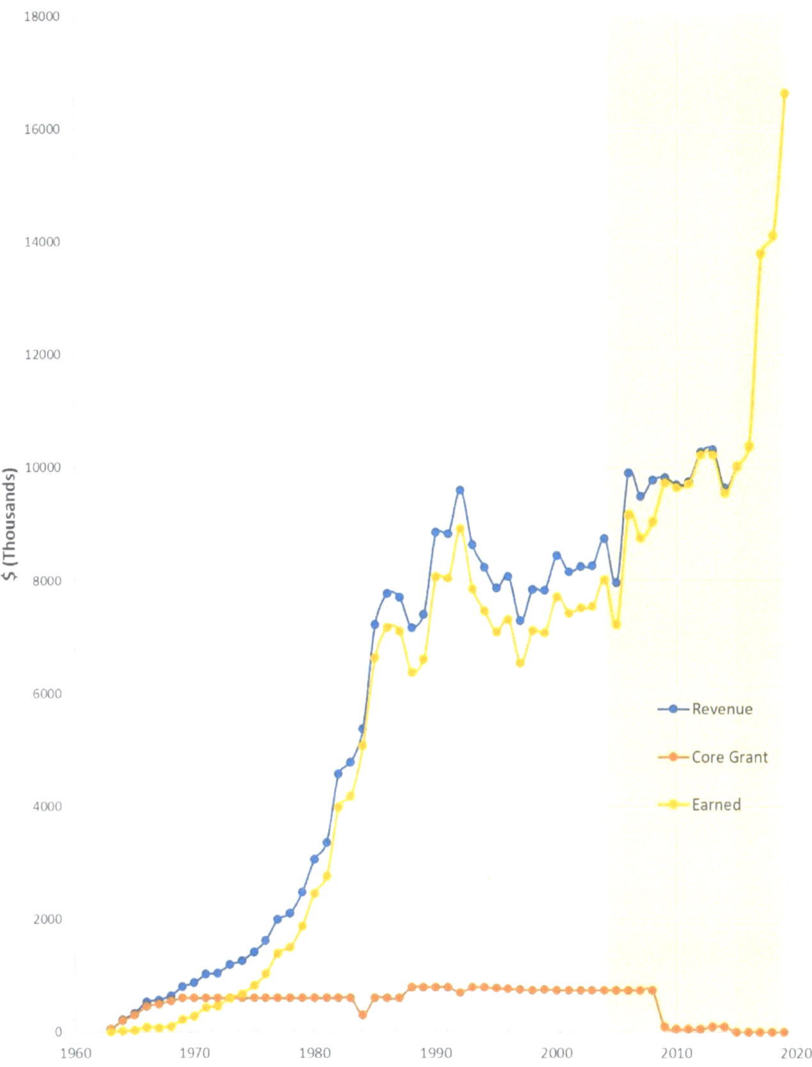

Figure 5.17. RPC revenues to 2019 (fiscal-year-end basis). Between 2014/15 and 2018/19, total revenue and earned revenue are the same. *The Entrepreneurial Years* **period is shaded.**

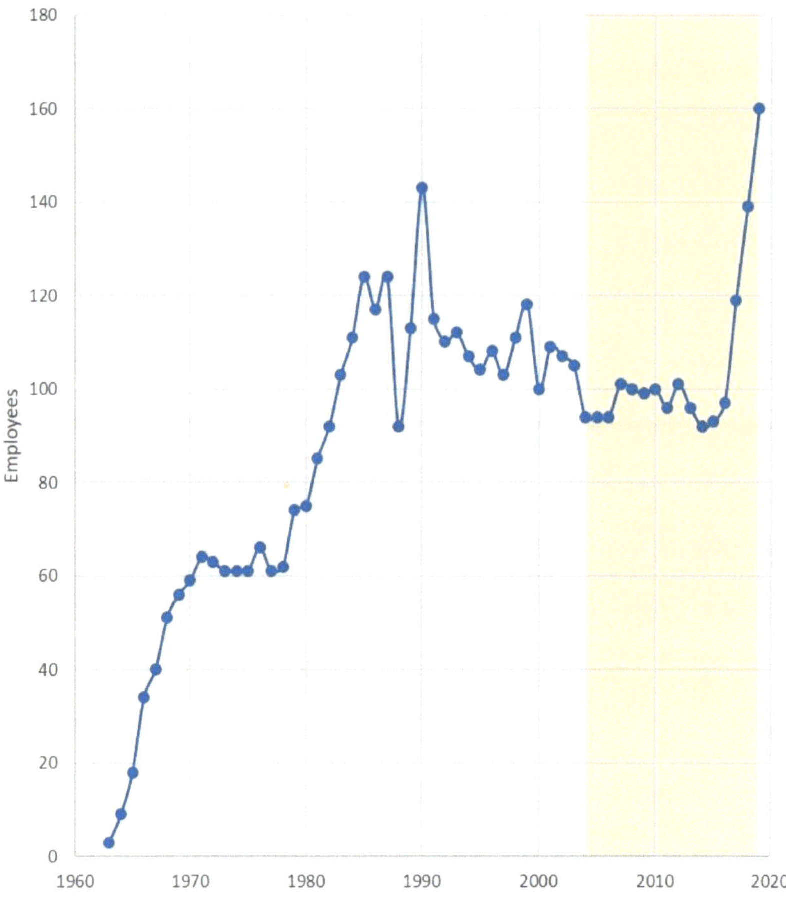

Figure 5.18. RPC employee levels to 2019 (fiscal-year-end basis). *The Entrepreneurial Years* period is shaded.

Table 5.1. Illustration of some of the changes experienced by RPC during *The Entrepreneurial Years, 2004 – 2020*. See also Tables 2.2, 3.2, and 4.1.

	2004	**2019**
Total Revenue	$8,746 k	$16,628 k
Provincial Grant	$741 k	$0
Earned Operating	$8,005 k	$16,628 k
Fixed/Capital Assets (mostly buildings, lab equipment and fixtures)	$2,528 k	$8,429 k
Employees	94	160
Clients Served	830	1127
Reference	[213]	[280]

6 THE IMPACTS OF RPC'S WORK

From its inception, RPC has had a vital interest in the economic impacts of its work. However, for the most part, these were only measured infrequently and, even then, only for specific projects or programs. Over the years, as RPC's industrial work matured, an increasing number of such economic impacts began to be reported by the clients themselves. Most of the impacts, and their associated stories, date from the 1970s and subsequent decades. The most commonly reported impacts were in new product developments and launches, product development cost savings, production cost savings, sales and export increases, job creation, and layoff/bankruptcy avoidance. By the 2000s, client impact reports broadened beyond economic impacts and jobs to also include social and environmental impacts.

In the following sections we have captured some of the impact data that is available, and a selection of the stories. It is the stories that paint perhaps the most vivid picture of the positive impacts of RPC's work over the decades.

6.1 RPC's Impacts in *The Building Years*, 1962 – 1969, and *The Growing Years*, 1970 – 1983.

In some years, RPC was able to estimate and report the impacts of its work for clients in terms of cost-savings, revenues, or profits. In 1967/68, for example, RPC's work achieved cost savings of "some $540,000" for numerous companies in NB and PEI [49], while in 1973/74, RPC's work in just three projects contributed to returns of some $120,000 for the three NB companies concerned [82].

In other years, individual RPC divisions (or departments) were able to report such estimates. For example:

- o In 1976/77, the work done by the Chemistry and Food Sciences Department contributed to estimated industrial revenue gains of approximately $10 million [124],
- o In 1977/78, the work done by the Engineering Department resulted in client savings of approximately $260,000 (an average benefit/cost ratio of 4.5:1), while work done by the Management Services Department enabled increased client revenues of approximately $1.3 million (an average benefit/cost ratio of 8 to 1),
- o In 1978/79 the work done by the Chemistry Department achieved a revenue increase or cost reduction of approximately $840,000 (an average benefit/cost ratio of 5 to 1) [138],
- o In 1979/80, a survey of twenty projects completed during the year by the Management Services Department revealed benefits of $2.7 million per year for the client companies, in terms of increased production or reduced operating costs [134], and
- o In 1982/83, the Engineering Services Department reported [145] one case in which material testing enabled a client to upgrade a shipment of low-cost material, avoid potentially hazardous material failure, and realize a benefit/cost ratio of 20 to 1, plus significant time savings.

Some RPC Impact Examples from 1978/79 [138]. A few examples of the jobs undertaken by RPC in 1978/79 illustrate the range of activities carried out and the benefits to RPC's clients. A major franchise chain had rejected product from the client's plant because of a 'medicinal taint' and was threatening cancellation of a purchasing contract which, in turn, meant probable plant-closure. Within three days RPC had traced the cause and source of the problem, and its report helped to persuade the purchaser that the firm had tackled the problem with utmost urgency and that future product would be satisfactory.

A nut processor was facing recall-of-product action from federal authorities because of alleged high residue content (aflatoxin). RPC's investigation showed the cause to be very occasional and unavoidable contamination of individual nuts where one such 'bad' nut would be sufficient to make a small specimen reject. When the nuts were correctly sampled all batches were found to be satisfactory and the product remained on sale.

Sometimes RPC's work results in immediate cost-savings. RPC's trials on an alternative way of making a new sauce mix for a food franchise chain showed how an original need for $120,000 of equipment could be reduced to $35,000 without sacrifice of capacity or quality. On yet another occasion, RPC's effort was part of a collaborative effort, saving a client from damage claims. In one such case RPC showed that chocolate bars could not be the cause of claimed illness and in another, that the quality deterioration in a $12,000 dispute on a nutmeg shipment was not due to mishandling by the client.

Finally, while the bulk of RPC's activities continued to be directly related to local problems, RPC was successful in obtaining a contract with the United Nations Industrial Development Organization (UNIDO) to study the best way of exploiting a major wild resource - the fruit of the tropical tree *Balanites aegyptiaca*. Meanwhile, two major projects were concerned with the expansion of the N.B. fishing and fish processing industries to take advantage of opportunities presented by Canada's 200-mile limit. A practical means of enabling the New Brunswick offshore fleet to take advantage of improving catch rates in the early spring was recommended. Projected benefits amounted to some $2.0 million annually. New export marketing opportunities for New Brunswick fish processors resulting from Canada's emergence as the leading exporter of fish products were also identified, and a plan was developed for providing infrastructure and management assistance to allow the industry to meet these market opportunities.

Some RPC Impact Examples from 1981/82 [123]. A few examples of the jobs undertaken by RPC in 1981/82 illustrate the range of activities carried out and the benefits to RPC's clients:

o Analysis and duplication of an imported fur cleaning compound created annual savings for a client of $10,000,

o Formulation of a corrosion inhibiting blend for heavy water plants enabled annual sales for a client of $250,000,

o A crab processing plant with production valued at approximately $150,000 per day was closed by Federal order. One day later, after an urgent inspection by an RPC bacteriologist and food technologist, the plant was allowed to re-open,

o Improvements made to a canned meat product to allow production by machine rather than by hand saved a client $35,000 per year,

o Development of a new canned product enabled a client to extend its market coverage and maintain its labour force,

o Improvements to the formulation and processing of a frozen meat product enabled a client to realize increased sales potential, while saving $40,000 per year in ingredients,

o Development of a novel sauce enabled a N.B. entrepreneur to begin business with an estimated first year profit of $25,000,

o A new chicken product was developed for a client, and made ready for test-marketing,

o Analysis and duplication of three spice mixes enabled a local processor to become independent of imported blends, o Nutritional analysis of a range of products enabled a client to satisfy mandatory labelling requirements for sale to the U.S., and

o Assistance to a fish packer with brining and pickling operations enabled it to maintain its European export markets.

6.2 RPC's Impacts in *The Commercial Years*, 1983 - 2004.

Some RPC Impact Examples from 1984/85 [149]. A few examples of the projects conducted by RPC's Food Science and Technology Department in 1984/85 illustrate the range of activities carried out and the benefits to RPC's clients:

o Assistance to six companies to improve their products,

o Formulation of nine new products for local companies including one for a PEI processor,

o Saving an estimated $50,000 per year in ingredient costs and at the same time giving the possibility of a new franchise-type venture,

o Investigating a potato storage fire and assisting in the salvage of approximately $200,000 of product,

o Acting as professional experts in a case involving disputed quality of an 18 tonne (20 ton) export shipment of smoked fish, and

o Analysis of poultry feed to discover the cause of growth defects and to prevent continuing production losses of about $15,000 per month.

Some RPC Impact Examples from 1988/89 [151]. Some examples of the projects conducted by RPC in 1988/89 include:

o In the *food sector*, assisting companies with problems in quality assurance, process design, food engineering, and ingredient analysis, plus developing ten new products for companies. For example, RPC helped a N.B. entrepreneur produce a non-frozen French fry for the fast-food industry, that maintained the product's desired colour and freshness while being transported and stored at room temperature,

o In the *agricultural sector*, working with N.B. potato-packing companies to identify and resolve problems related to tuber quality during packing operations,

o In the *fish processing sector*, working with two major N.B. companies on the development of value-added herring-roe products and improving the quality of frozen product prepared for overseas export,

o In the *aquaculture sector*, developing a rapid, simple and inexpensive kit method for early detection of serious diseases in farmed Atlantic Salmon, and

o In the ***manufacturing sector***, RPC helped a local metal fabricator and a manufacturer develop a machine that can disassemble and reassemble the track components on military tanks. This allowed the companies to fulfil a major contract with the Department of National Defense [281].

Skunked: An RPC Product Development Example *Circa* 1989. In the late 1980s, RPC employee Charles Wiesner developed a product to remove the odour of skunk spray from pets. According to a *New York Times* article [282]:

"the odor in skunk spray comes primarily from two classes of highly sulfurous compounds. Both can be changed by linking the sulfur to oxygen molecules. Mr. Wiesner developed a series of chemicals that can do the job, his favorite being lithium iodate. He has incorporated the chemicals into a shampoo that can wash the skunk spray out of fur, hair and clothing."

The skunk-spray shampoo was patented in Canada and the U.S. (see the Wiesner patents in Appendix 8.8), and licensed to Canadian Odor Control Ltd. (Brighton, Ontario), for manufacture and sale under the name 'Skunked.' This product is still available in the marketplace at the present time and is reported to be carried by many veterinarians in New Brunswick.

An RPC Impact Example from 1997 [226]. As noted in Chapter 4, an unknown virus had been causing Haemorrhagic Kidney Syndrome (HKS) and killing Bay of Fundy salmon [169,170]. RPC scientists were able to identify the virus that was causing the HKS and the deaths: Infectious Salmon Anaemia (ISA) [171]. Armed with the results of RPC's work, industry and government were able to take steps, including reducing the amount of effluent from fish-processing plants, to prevent a recurrence of the disease [170]. The impact of this work was the saving of a $120 million-per-year industry in New Brunswick.

Additional protection was developed in subsequent work, through which RPC's Dr. Steve Griffiths and his team discovered a naturally-occurring bacteria, *Arthrobacter davidanieli*[57], that has a similar makeup to BKD but when injected into fish helps them to develop an immune response [173]. The vaccine was licensed to Novartis for sales under its commercial name, Renogen®, and it is currently registered to and manufactured by Elanco for sale in Canada and worldwide.

Some RPC Product Development Impact Examples. Some examples of product development projects completed by RPC by 1994/95 [283] and 1995/96 [284] include:

o RPC helped F.A.H.R. Industries (Edmundston, N.B.) to develop their Forester C-2000 rock crusher for use in surface conditioning of forest and other wilderness roads [283].

o RPC helped Transform Pack Inc. (Moncton, N.B.) develop TRANSFORM Pack™, a novel meat- and seafood-packing packaging technology in which the packaging film contains spices and ingredients that are 'unlocked' by moisture in the meat, so that curing or marinating can begin immediately under controlled conditions making sure the product ready to cook when removed from the refrigerator [283]. (See also the patents by Meier *et al.* in Appendix 8.8 and the section on "A Client Update" (2001/02) below.)

o RPC helped Stolt Sea Farm Inc. (Saint George, N.B.) to develop a value-added product line that included Fresh Marinated Salmon Kabobs [283].

---

[57] Named after Dr. Griffith's son, David Daniel Griffiths.

o RPC developed a novel three-speed wheelchair hub (to enable wheelchairs to better negotiate access ramps and thickly carpeted surfaces), for a client in Dalhousie, N.B. By 1994/95, this patented product had been thoroughly tested by a wheelchair manufacturer preparatory to manufacturing retrofit kits in New Brunswick that would fit on popular wheelchair models [282]. (See also the patent by Godin and Lambert in Appendix 8.8.)

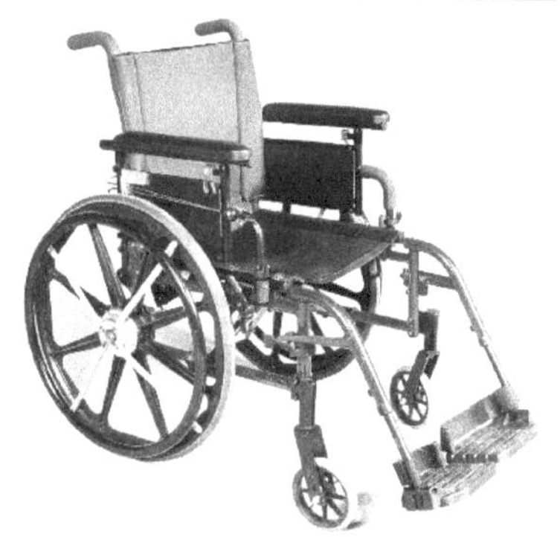

o RPC helped an entrepreneur from Northern New Brunswick create a contoured broom-ball paddle to allow ball control analogous to the effect achieved by a curved hockey-stick blade [284].

o RPC worked with a local veterinarian to develop a unique DOGVAC™ product for grooming pets [284]. In this case, RPC assisted with all aspects of the product development process, from concept evaluation through to working prototype and design of production molds.

Some RPC Client Impact Testimonials from *The Commercial Years.* It has been said that *"The voice that can credibly speak to the impact(s) of R&D is the voice of the customer"* [285].

"I believe this project would never have gotten off the ground without RPC. RPC took us under their wing, guided us through the initial stages when we needed help and have been there for us throughout the whole process."

A successful New Brunswick entrepreneur,
Quoted in RPC's Annual Report 1995/96 [284].

Some examples of RPC client voices from this era include the following.

Mynic Inc. Following graduation from UNB's forest engineering program, New Brunswick entrepreneur Mike McCormick launched Mynic Inc. to pursue the development of a commercial log-scaling system (a measuring device to measure the diameter of cut logs). Having achieved success with his first two product generations (Mynic and Mynic II), he began working closely with RPC on further developments in 1987. Access to RPC's experienced engineers and state-of-the-art computers led to the development of Mynic III and, later, Mynic IV.

"RPC provided the expertise to make our product reliable," said McCormick. RPC also helped develop a new computer circuit board, design a barcode-reader input system with a flexible scale-stick linked to a high-precision caliper (to improve measuring accuracy and enable the measurement of seedlings). Working together, *"we enhanced the Mynic III's reliability by 500 percent. Mynic IV has more calculating capability, a longer battery life, faster assembly time, is smaller and weighs less. It's also capable of working in colder temperatures (-40 C)"* reported McCormick, *"Mynic IV can record any data required by forestry or any other industry. It can calculate trigonometry, provide statistics and accommodate note taking."*

By 1990, Mynic sales had spanned pulp and paper plants, forestry contractors and consultants, government and training institutions, and saw-mills, and the company was looking to export its product to the rest of Canada and the U.S. According to McCormick, *"RPC has given us the means to provide the forest industry with a dependable and efficient data collection system. That was and still is the goal of Mynic."* (Original story published in 1990 [204].)

Paul's Cookie Company. One day, entrepreneur Paul Coté made up a large batch of cookies and gave them away to friends spreading the word that he was "in the cookie business." His friends took him seriously. The next day he had orders for 37 dozen. In the fall of 1987, Paul's Cookie Company officially opened for business, beginning with three original family recipes -

chocolate chip, peanut butter and oatmeal chocolate chip. His first retail outlets were a local farmer's market and a local meat market. Before long, he needed to increase his recipe batch portions and move to a new facility. Soon, he was serving approximately 250 retail outlets throughout central New Brunswick, with production rising to 10,000 cookies per day.

When demand grew to the point where the recipes needed to be modified once again to accommodate larger volumes, Paul approached RPC for assistance in scaling-up production while retaining the original flavour and texture. RPC helped with product development (improving and upscaling the recipe) and improving the production process.

"RPC has been a tremendous resource to my operation," said Paul in 1990. *"I only wish I had gone to them sooner. I would recommend their valuable research and development services to any manufacturer… Thanks to the assistance of many friends and RPC, production has increased by almost 50 percent. It's certainly been a successful three years."* (Original story published in 1990 [204].)

Aqua Health Ltd. Launches the World's First Approved BKD Vaccine. When Aqua Health Ltd. received a Canadian license for a Bacterial Kidney Disease vaccine for use in salmonid species in 2000, the Canadian Food Inspection Agency became the first regulatory agency in the world to approve this valuable aid in the prevention of BKD.

Marketed under the trade name Renogen™, this unique live-vaccine product had repetitively proven its safety and ability to minimize the difficulties caused by *Renibacterium salmoninarum* in the laboratory and during extensive field trials. The vaccine is the culmination of four years of research performed in conjunction with RPC. Researchers have learned that the traditional whole cell killed vaccines do not provide sufficient protection against *Renibacterium salmoninarum* therefore other solutions had to be investigated. In 2000, Renogen™ was the only approved live-injectable fish vaccine in the world. (Aqua Health press release, 2000, referenced in [173].)

Titan Trials BIOHEAP™ with INCO Ltd. In 2001, Titan Resources NL announced it was conducting scientific trials with its proprietary BioHeap™ bacterial leaching technology for one of the world's leading nickel producers, Inco Ltd. The metallurgical work with Inco followed excellent laboratory test results and was to involve large-scale column leaching tests on ore samples from one of Inco's Canadian nickel deposits. The tests were planned to be conducted in a field environment at the facilities of RPC. RPC had been, and continued to be, closely associated with the development of BioHeap™. Aimed at the successful implementation of bacterial bio-leaching under conditions of extreme cold, the trials were expected to serve to widen the scope of the company's BioHeap™ development program. (Titan Resources NL press release, 2001, referenced in [175].)

Waypak Finds Markets Outside Its Own Backyard. Airline absorber mats are very much like large diapers which are laid down beneath shipments of perishable products that may leak or spill during shipment. The mat will absorb and contain the liquid during the trip. At the final destination, there is no concern for clean up, and the ground crew can unload the shipment efficiently and without any delays associated with spillage.

By 2000, Waypak had held the national supply contract with Air Canada for several years and had been awarded the contract for another two years. The machinery for this unique application was custom designed and built for Waypak by RPC in Fredericton, NB. (Business East, 2000, referenced in [175].)

RPC's ***Innovative Nozzle System Inspection Technique*** provides an example how the economic impacts of RPC's work for a New Brunswick industry (in this case, nuclear power generation) can also be created beyond the province. One of the outcomes of RPC's work for the CANDU Owner's Group (*circa* 1989/90) was an ultrasonic inspection probe technology. RPC's ultrasonic probe (Figure 6.1) could be inserted into the reactor core to measure the proximity of calandria[58] tubes to horizontal-flux-detector tubes, which was needed to calculate fuel channel sag and pressure tube distortion [286].

RPC later incorporated this technology into other special ultrasonic inspection tools for use in nuclear reactors (*circa* 1991-92) [166]. RPC and Ontario Hydro developed a system for inserting an ultrasonic inspection probe, from outside of containment, and measuring the gaps between calandria tubes and shutdown system nozzles. Such gaps need to be maintained to prevent the calandria tubes from contacting and damaging the nozzles. Whereas the utility's plans originally called for the nozzles in Bruce Units 1,3 & 4 to be replaced in 1995-96, the measurements conducted showed that the actual gaps were larger than predicted by the utility's computer models, so much so that the nozzles might not need replacement for the entire life of the reactors. Based on this work, by postponing or eliminating the need to replace the nozzles, Ontario Hydro Nuclear estimated that they would save as much as $8-million over the operating lifetime of the reactors [287].

---

[58] In a CANDU nuclear reactor, 'calandria tubes' house pressure tubes containing fuel and coolant within the cylindrical reactor vessel, which is termed the 'calandria.'

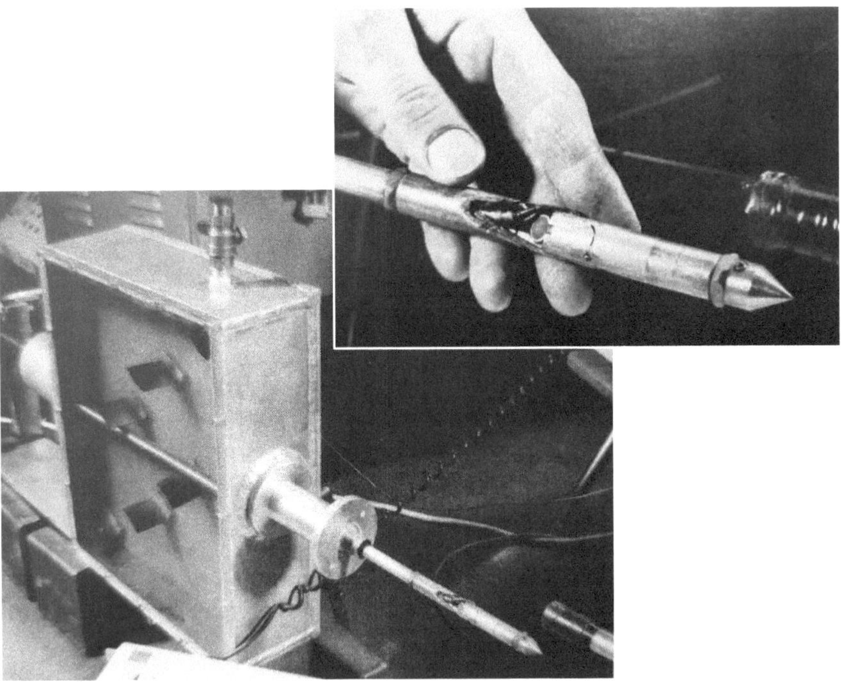

Figure 6.1. A prototype of RPC's ultrasonic monitoring tool [288] (Lower-Left). A close-up of the ultrasonic probe itself, showing the transducers and guiding bearings, is shown in the upper-right [286]. RPC photos 1989, 1991.

In the 2000s, RPC began to more systematically identify and report examples of its positive impacts. The following examples have been slightly edited, for consistency, from the originally published stories.

Development of a Strain Typing Assay for ISAV (2001/02, [241]). In 1997, *infectious salmonid anemia* virus (ISAV) was first identified within aquaculture cages of the Bay of Fundy. The initial focus on surveillance and fallowing of infected sites had expanded to include characterization of ISAV isolates within Atlantic Canada so as to identity strains that might be associated with high host mortality and to enable industry to track the disease. To achieve this characterization, RPC developed a diagnostic assay for use in surveillance programs that is capable of providing quick, reliable results for ISAV detection and concurrently capable of typing ISAV isolates [289]. In this assay, a combination of Reverse Transcriptase polymerase chain reaction (RT-PCR), which has proven useful in surveillance efforts, with Denaturing Gradient Gel Electrophoresis (DGGE) technology was used to identify and type ISAV based on nucleotide sequence variability within a specific gene segment. Following initial standardization with known isolates, DGGE may be performed on RT-PCR products to rapidly determine the strain of each sample without the need for additional manipulations such as sequencing or restriction enzyme digestion.

As a follow-up to the above story, in 2014, RPC's Dr. Ben Forward was able to provide the Standing Senate Committee on Fisheries and Oceans with a perspective on the next steps that were taken [174]:

"Once identified, diagnostic tool development ensued and allowed for the monitoring of this virus to control its spread and avoid costly economic losses. Subsequent work then focused on the development and implementation of strain typing tools and disease challenge models to help us identify and characterize the virulence of the different viral strains we found out in the environment. This is a critical development which allowed regulators and farmers to make informed decisions regarding the health of their fish."

Pioneering New Inspection Technology (2001/02, [241]). RPC work advanced an emerging technology – Guided Wave Ultrasonics - in order to carry out critical inspections for the CANDU nuclear industry. Guided waves are those whose propagation characteristics depend on structural boundaries. Internal inspection of tubing had been traditionally accomplished using a rotating transducer to scan the tube wall point by point. With guided waves, a wave can be launched from a single location that will propagate several metres along the tube yielding echoes at crack locations. It is even possible to segment the receiver element so that the circumferential location of the reflector (crack) can be determined. As a result, guided wave inspection had potential for application where internal tube access is restricted or denied.

RPC adapted this technology to develop a guided wave technique specifically for internal inspection of outlet feeder bends in CANDU reactors, looking for cracks in the tube wall. Testing the method and training inspectors presented its own challenges. RPC designed and manufactured calibration tubes with miniature defects machined (by plunge electro-discharge machining) into the tube wall on the internal bend surfaces. Because the guided wave technology identifies the circumferential location only, RPC developed other ultrasonic technology to subsequently detect the length of cracks.

RPC's 'TIP' technology uses new and innovative software to reduce noise from conventional crack tip ultrasonic signals so that the location of the tip can be determined. The length of the crack is determined by locating the partner tip. Cracks in outlet feeder bends impose limits on the service life of nuclear plants. One hundred percent inspection of outlet feeder bends, using RPC's Guided Wave and TIP technologies, was scheduled for the maintenance outage of Pt. Lepreau Nuclear Generating Station in June, 2002.

A Client Update (2001/02, [241]). Eight years prior to 2001/02, RPC worked with a start-up firm in the food industry to develop a novel packaging film. (The project was featured in the 1994/95 RPC Annual Report [283]). The film was impregnated with spices and ingredients and used to wrap frozen meat and seafood products. When the product is thawed in a refrigerator, the ingredients in the film are "unlocked" by the moisture in the meat so that curing or marinating can begin immediately under controlled conditions. The product is ready to cook when removed from the refrigerator, reducing subsequent preparation time. The market niche envisaged upscale, rushed consumers with a taste for gourmet food.

The Atlantic Provinces Economic Council in its **ATLANTIC** *Report* magazine (Winter 2002) reported that a Moncton firm, Transform Pack International Inc. (RPC's client of eight years previously), was to provide Tilia Inc., of San Francisco, with a minimum of $10.5 million worth of its patented seasoning sheets, which marinate and spice meats. Contacted by RPC, the company confirmed it was doing well, had no competitors for its patented product and had markets in Europe, too. A talent for innovation, industrial partners, nurturing scientists and technologists for success in New Brunswick over the long term - all "ingredients" in this case history.

The Right Stuff at the Right Time (2002/03, [3]). In the 2001/02 Annual Report (p. 8, [241]) RPC reported that they had purchased an advanced LC/MS-MS[59] instrument because of growing concern over drug residues and carcinogens that are present at trace levels in the environment. RPC had anticipated that some N.B. firms would soon have a need for this technology. On April 24, 2002 the international news media reported that a study carried out by Stockholm University and scientists at Sweden's National Food Administration showed that heating of carbohydrate-rich foods - potatoes, rice, cereal - formed acrylamide, a substance classified as a possible human carcinogen.

Sure enough, a manufacturer in New Brunswick immediately began planning its own test program to investigate ways of reducing levels of acrylamide in processed food products. The manufacturer called RPC to inquire if they could analyze for acrylamide at the trace levels cited in the Swedish study (50 parts per billion). RPC confirmed they could assist, using the newly acquired LC/MS-MS technology. Soon RPC was analyzing samples from the client's test program on a five (sometimes three!) day turnaround, and to within 50 ppb - all on a strictly confidential basis.

---

[59] Liquid chromatography with tandem mass spectrometry is an analytical technique that combines liquid chromatography (for separating chemical species) with triple-quadrupole mass spectrometry (for selectively analyzing the separated species).

By year-end, the client's test program had been completed and the client granted permission to lift the veil of secrecy so that RPC could advertise acrylamide analysis, inform other potential clients of the newly acquired capability, and demonstrate publicly how highly-qualified people, state-of-the-art technology, and prudent allocation of capital resources all help to keep RPC ready to serve industry leaders in New Brunswick.

Whose Moose is Loose? Find out with Wildlife DNA Forensics... (2002/03, [3]). RPC had developed a moose genotyping assay helpful to wildlife managers and law enforcement. In 2002/03, RPC sought accreditation of its forensics laboratory, for moose and other species, so that findings from evidence obtained by law enforcement would stand up in court. Previously, wildlife law enforcement had only been aided by techniques for species identification of unknown forensic samples. However, by 2002/03, it had become possible to determine the DNA profile of an individual animal. This information, in conjunction with a relevant population database of DNA profiles, could permit determination of the frequency of any given genotype and calculation of the statistical probability of a match between two samples. New Brunswick did not have a database of such profiles, creating the need for RPC to get involved.

RPC's work included developing a simple, fast, and rigorous assay for genotyping moose samples, establishing a DNA database containing genotype data from a sample of the N.B. moose population, and building a framework for discrimination between two moose samples for use in prosecution and deterrence of illegal activity. Samples were collected from high- and low-density populations and from all major geographic regions of the province and the DNA was extracted. A multiplex assay was developed that can determine a genotype from moose tissue within 48-72 hours. A population database of moose in New Brunswick was established using 120 samples from across the province. The database provides a good

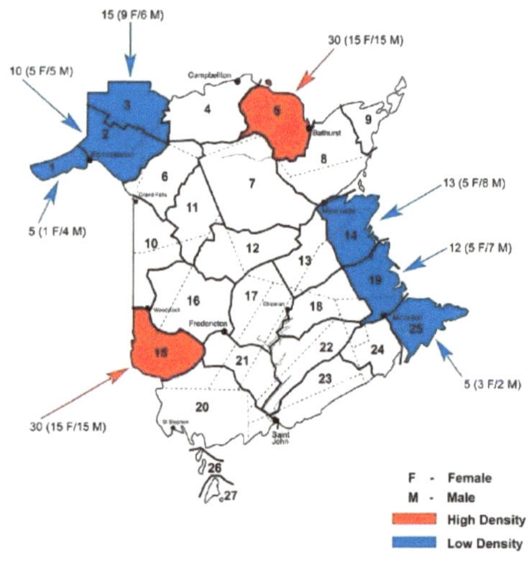

representation of the New Brunswick moose population. See also "CSI Deep Woods," below.

A Global Marketplace for RPC's Expertise (2002/03, [3]). When RPC was awarded a second project under the Canadian International Development Agency's Canada-Southern Cone Technology Transfer Fund, it had to draw upon its technical expertise in the areas of laboratory management, environmental analytical services and pesticide application technology. The focus of the previous project had been the transfer of Canadian technology to improve the use and management of pesticides in Chilean agriculture.

In the 2002/03 project, RPC coordinated a group of 20 Canadian public and private sector organizations and their South American counterparts to extend the program to two additional countries while building on the results achieved in Chile. In Uruguay, RPC would assist the Ministry of Agriculture and its partners to develop improved pesticide management legislation, including standards related to the management of empty pesticide containers, its capacity to better evaluate the risks related to pesticide use and their impact on public health as well as to develop educational programs directed towards users of agricultural pesticides.

In Paraguay, the project would focus primarily on the development and delivery of educational programs for small agricultural producers. In Chile, RPC would continue work to reinforce the technical, analytical, monitoring and inspection capabilities of agencies that are mandated to regulate the use of pesticides to improve safety in their application and use. Pesticides are widely used in South American agriculture to protect the export value of key agricultural products. The project was due to be completed in 2006 and had a total estimated value of $4.3 million.

Stealing Away from the Crime Scene... (2003/04, [213]). The popular TV show Crime Scene Investigation (CSI) is fiction, but the forensic science it shows is real. RPC is using the same DNA technology to find the genetic 'fingerprints' of fish, a process called 'genotyping'. Genotyping provides information to New Brunswick's $200 million aquaculture industry needs to improve its breeding programs, and to understand how to sustain healthy, genetically diverse stocks. Until just a few years prior, conventional breeding methods were used to choose the best fish to be bred to produce strong and healthy offspring. The fish to be bred, called 'broodstock,' were tracked by physical branding or tagging and their success was measured throughout their lifetime. This process goes a long way toward developing healthy fish for farmers, but more could be done.

Just like any other animal, fish need to be bred carefully to avoid unhealthy inbreeding. The emergence of genotyping brought farmers a tool for assessment of the genetic diversity of their stock. If inbreeding is detected, farmers can improve the genetic diversity of their fish stocks by setting up mating between two broodstock fish with different DNA genotypes. By maintaining and enhancing genetic diversity of broodstock, fish are healthier and grow bigger and stronger.

Genotyping is also important to what is called 'traceability,' the ability to trace the origin of a fish from the supermarket back to the processing line and further back to the fish farm. Also, genotyping is a handy weapon to have in dealing with trade restrictions encountered when Canadian fish farmers export live fish. The salmon species native to several salmon rivers in the State of Maine had been declared endangered, and therefore had to be protected from breeding with 'undesirable' fish, such as escapees from aquaculture cages. A law was passed that all salmon placed in aquaculture cages in US waters had to be shown to be of North American origin, just in case they were to escape and engage the locals.

RPC's genotyping service allows Canada's exported fish to be screened against a large database of genotyping information from fish in Europe and North America to verify its continent of origin. In this case, threats to endangered species and adaptation to international trade restrictions were issues RPC could solve with CSI-like DNA technology.

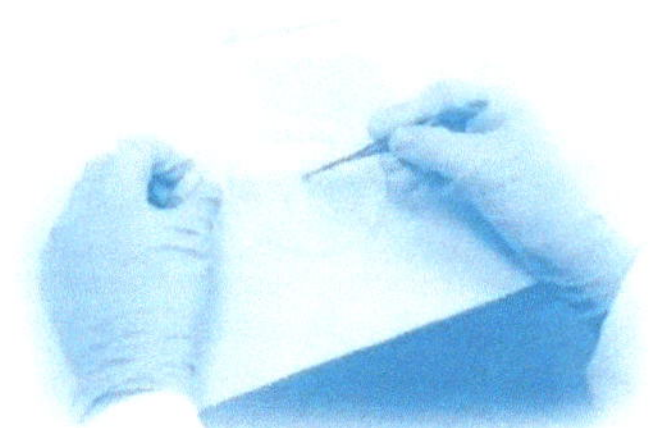

The Clue was that One-Billionth of a Gram of Lanthanide... (2003/04, [213]). Otoliths ("earstones") are small, white structures found in the head of all fishes other than sharks, rays and lampreys. Otoliths provide a sense of balance to fish in much the same way that the inner ear provides balance in humans. Fish otoliths also aid in hearing, and are important in fisheries research. One aspect of otolith investigation, involves trace element analysis or 'elemental fingerprinting.' Researchers have linked otolith chemistry to otolith biochronology, and the elemental fingerprint provides a natural 'tag' of fish stocks. From an analytical chemistry standpoint, however, the elemental analysis of otoliths presents a number of challenges as most of the trace elements of interest are present at very low concentrations. Special cleaning and handling techniques are essential along with superbly sensitive analytical instrumentation, a contamination-controlled work environment and ultra-pure chemical reagents. In order to detect subtle differences in trace element ratios among otoliths from different fish populations, high-precision analysis capability is mandatory.

RPC elected to use an inductively-coupled-plasma mass-spectrometer (ICP-MS) for conventional analysis, and 'isotope dilution' analysis, of eight targeted trace elements in some qualification tests. After successful completion of the initial method development RPC staff proceeded to analyze 1,700 specimens in 1997/98. In 2002/03, they completed two separate contracts for the U.S. Environmental Protection Agency. For this work, the team also developed methodology to accommodate the very small samples involved, some weighing less than 0.001 grams. The total quantity of a particular trace element present in such a small sample could be in the billionths of a gram range. For this, they developed micro-sampled ICP-MS techniques.

In 2003/04, RPC acquired a new ICP-MS that is capable of better sensitivity and stability, and began construction of a clean-laboratory sample-preparation area. With this facility, RPC has been able to confidently report more trace elements in otoliths, at even lower concentration ranges, and has used the new equipment for the successful completion of additional contracts in support of researchers from Canada, the U.S., and abroad. Subsequent work involved

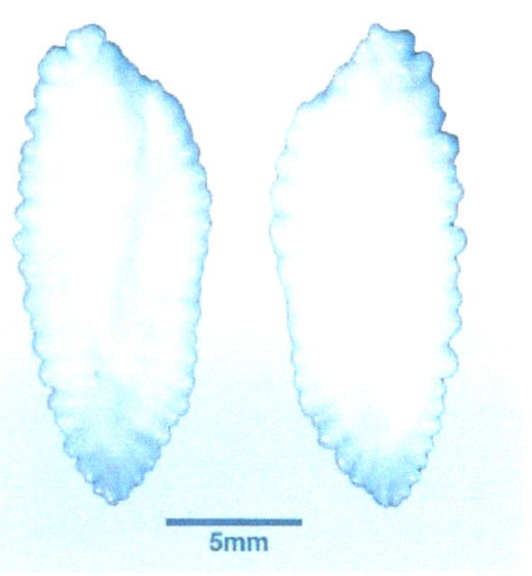

further development of methods for the determination of ultra-trace lanthanide concentrations, and other ways to enhance precision, accuracy, and sensitivity of the analyses to better meet the needs of RPC's clients in this demanding and complex area.

Is there a hemi in there? (2003/04, [213]). Trucks perform work in moving objects from point to point. Aficionados say that trucks are not all alike - they vary in their capacity to perform work, on a power per unit weight basis. Indeed, a given model may change from year to year with respect to its horsepower per pound capacity in response to a signal from consumers.

Enzymatic proteins can be like trucks. These combo fuel/engine/pump units are integral within the cell walls of living organisms. They perform work by making a conformational change to move a solute ion through the cell wall. Specifically, the Na^+/K^+- ATPase pump transfers sodium ions out through the cell wall, and transfers potassium ions back in, using energy from a fuel source - adenosine triphosphate. These pumps are crucial to the cell because they remove and replace ions that leak in and leak out of the cell respectively, thus maintaining the ionic balance across the cell wall. Maintaining the ionic balance across the cell wall is also vital to balancing the osmotic pressure on either side of the cell wall, or water could alternatively flow into or out of the cell causing its destruction.

When fish move from the fresh water of a hatchery or stream to the salt water of an aquaculture site or the open ocean shortly after the parr-smolt transformation, the fish and its cells are exposed to huge changes in osmolarity. The activity (power per unit weight of protein) of the enzymatic protein pumps in the gill cell walls of the young fish must be sufficient by that time to meet the increased demand for work to balance the salt concentrations on either side of the cell wall. In fact, the power of the pumps in gill cells begins to increase during the parr-smolt transformation, prior to transfer to salt water, in response to a signal (the appearance of a hormone).

One of RPC's clients was a supplier of special feed ration systems that promote a uniform parr-smolt transformation within a fish population so that the fish are all physiologically ready for salt water at the same time. To assist the client in applying the technology to N.B.'s aquaculture industry, RPC developed a method of measuring the activity of the pumps in gill cells and determining whether the fish

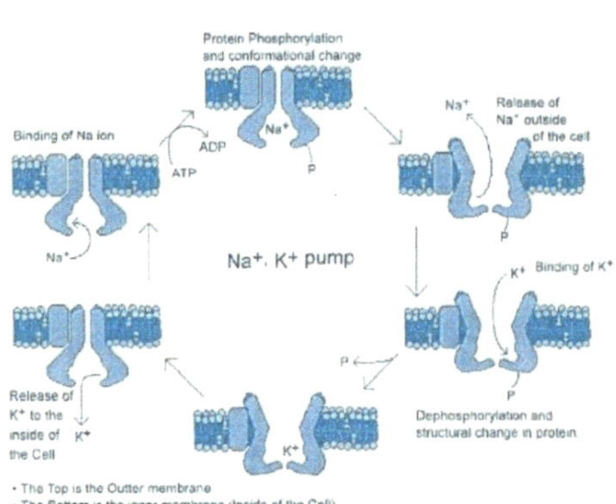

are physiologically ready for the move to salt water - too early can cause excessive mortality (pumps too weak), too late can cause excessive grilse production (over-production of signal hormones.).

One could say RPC tests that the right power level is achieved at the right time for the enzymatic proteins in the fishes' gill cells to cope with a greater workload.

Saving $20k per Day is Watt It's About... (2003/04, [213]). A nuclear power plant's operating limit is tied to its thermal power production. The measured value of feedwater flow rate is used in the estimation of reactor thermal power. The flowmeters commonly used contact the flow medium and accumulate corrosion products over time causing them to drift upwards in reading. Consequently, the reactor's measured thermal power is overestimated. To stay within regulatory limits, reactor operators are required by the regulating authority to de-rate their plants (from the thermal power output indicated by the overestimate) by as much as three percent of full power. More accurate flow information would provide a better estimate of reactor thermal power and would enable the plant to confidently operate closer to full power. The cost of the de-rate is typically $20k per day of foregone power generation for a 600 MW plant.

RPC had been working to perfect a non-intrusive ultrasonic flowmeter over the previous decade. Key advances included the development of transducers that are stable at temperatures up to 550 C, the emergence of powerful computers that have enabled progressive refinement of the measurement of the time difference between upstream and downstream propagation times of ultrasound through the pipe walls and flow medium, and advances in signal processing technology.

In 2003/04, RPC staff took the final step and had the flowmeter calibrated in a world-class calibration facility at Alden Labs in the U.S. The instrument was clamped on the outside of a 20-inch diameter pipe. Water was pumped through the pipe at a rate of about 5,000 USGPM and collected in a large tank on a weigh scale. The RPC flowmeter produced a new level of accuracy with results of 0.23% of actual flow. This level of accuracy is well within that required for the intended application in the nuclear power industry. RPC sold two flowmeters in 2003/04, to Japan Nuclear Cycle and GE Panametrics.

6.3 RPC's Impacts in *The Entrepreneurial Years*, 2004 - 2020.

A Fish Story (2004/05, [242]). You are what you eat. This fact has led to increasing concern about what is in the food you eat. The Michigan Fish Advisory is a database containing information on contaminants found in fish for locations throughout the state of Michigan. It is part of the Michigan Department of Environmental Quality's (MDEQ) comprehensive water quality monitoring program, an effort that has produced several years of data regarding contaminants. MDEQ required

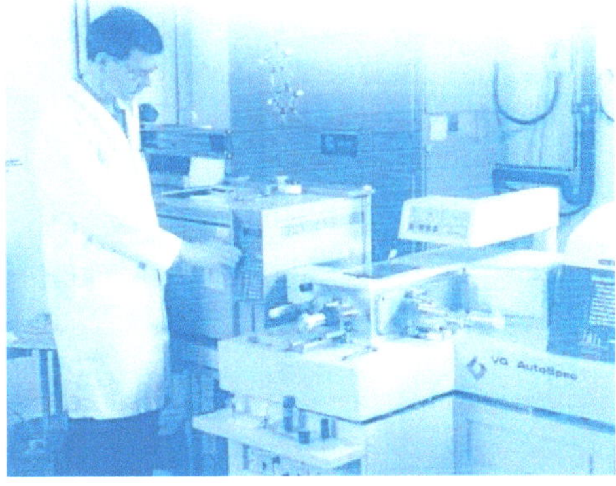

analytical lab services for more than a dozen environment pollutants and dioxins, including PCBs, DDT and mercury. Hundreds of samples are collected each year for analysis, and MDEQ called on RPC's expertise for the analysis of these samples.

RPC had been providing the lab services for this project for several years and was continually improving its service. In fact, efficiency improvements in 2004/05 resulted in a 75% reduction in processing time. The results of the program were used to assess chemical contamination in fish from Michigan's surface waters. Fish contamination data are used to determine whether fish are safe for human and wildlife consumption and to evaluate the effectiveness of DEQ initiatives to reduce contaminants. In this case, it is the fish telling the story about the environment.

***"If you can't measure it, you can't manage it."* RPC Determines the Facts About NB's Environment** (2005/06, [245]). In 2005/06, significant environmental issues receiving public attention in New Brunswick led to a need for RPC to help environmental consultants. In the Belledune area, several studies were conducted to investigate concerns about emissions from the smelter in the area. Working for multiple clients, RPC analyzed samples of soil, water, vegetation and tissue to determine the levels of lead, cadmium and arsenic, all of which are of health concern. By providing facts on the issue, appropriate actions and strategies could be identified and, if necessary, their effectiveness monitored. At Gagetown, concerns about Agent Orange and other chemical compounds provoked a study to investigate the possible presence of dangerous pollutants. RPC's unique capability in dioxin analysis helped it to win a national contract for analyzing soil, water, and vegetation samples. The analyses RPC completed to support the consultants' studies resulted in the availability of facts for the stakeholders.

Bringing Home the Bacon... from Chile! (2005/06, [245]). The Chilean pork industry exports its product to Europe, which has stringent, ever-tightening requirements for products to be certified dioxin-free. Dioxins can be highly toxic and persistent, so very low-level detection limits are necessary. Such limits require highly specialized 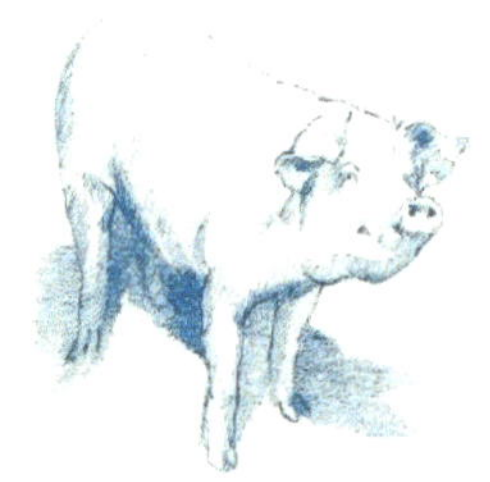 equipment, procedures, and personnel to conduct tests to accredited standards. When the Chilean pork producers needed help in monitoring dioxin levels, they turned to RPC to provide it. The clients in Chile froze samples of pork flesh and couriered them to RPC, where the dioxin levels were measured and reported back to Chile. Once the product was verified to be compliant with import regulations, the pork was shipped to European Union countries.

The combined effects of environmental contaminants and the globalization of the food chain places increasing emphasis on the importance of food safety. RPC is well positioned to apply its science capability to help with these challenges.

While RPC's services are recognized and acquired internationally, it regularly works with local food and beverage processes to address their needs for dioxin analysis. In the case of Chilean pork exports, the piggy goes to RPC before it goes to market.

Energizer: Nuclear Energy Safety. *Development and Installation of HEP and ECP Corrosion Monitoring Systems for Point Lepreau* **Nuclear Generating Station** (2005/06, [245]). As part of its Atlantic Innovation Fund partnership with the Centre for Nuclear Energy Research (CNER), RPC developed, tested, and manufactured two corrosion monitoring systems for use in CANDU nuclear power stations. The Hydrogen Effusion Probe (HEP) provides an on-line measurement of the rate of wall thinning in carbon steel pipes. It works by measuring the quantity of hydrogen which effuses through the pipe wall. The Electrochemical Potential Probe (ECP) provides an indication of the state of the oxide layer coating, the inside of the pipe and its susceptibility to the initiation of cracking. RPC staff worked with CNER to design and build the probes, prepared all the documents for installation on-site, built practice mock-ups, tested all the hardware, and wrote and tested all the software and electronics. This project provided an excellent example of collaborative research and development: the client was impressed with the rapid and timely developments, and the installation of both systems took place in 2006, at Point Lepreau.

For Sale: 'High' Quality Housing (2005/06, [245]). RPC is frequently called in to conduct inspections of homes about to be purchased or sold, or when there is concern provoked by an illness, or the sighting of visible mould. One of the interesting related projects from 2005/06 was the inspection and testing of former cannabis grow houses in the Moncton area. The RCMP had seized the houses and all the plants had been removed prior to the visits. The houses were inspected for any visible mould and testing was conducted for airborne spores. Bulk samples and swabs were also collected where visible mould was found. The concrete floors were also tested for herbicides and pesticides. Benzothiazole, a fungicide and an anti-microbial agent was detected in most of the houses. The compounds dronabinol and cannabinol (narcotics) were also detected indicating cannabis plants had been present in the houses. Once all the mould was removed and the floors cleaned, second visits were made to verify that the clean-ups were successful and the air quality was acceptable prior to the houses going on the market.

Keeping the Health in our Health Care System (2006/07, [246]). Establishing and maintaining an effective, safe health care infrastructure is essential and RPC has played a supporting role in this endeavor. For example, RPC provides essential medical gas certifications for all new and renovated medical gas piping systems in hospitals and other health care facilities. RPC is recognized as an accredited testing laboratory by the Standards Council of Canada (SCC) for the inspection of non-flammable medical gas piping systems and analysis of compressed breathing for compliance with applicable CSA standards. RPC's effective and responsive service established the organization as the preferred supplier for accredited inspection services in hospitals throughout Atlantic Canada, completing virtually all the inspections in the region, including analyses of medical gases.

During 2006/07, RPC supported major projects such as the new construction of the Upper River Valley Hospital in Waterville, as well as major expansions and renovations such as those at The Moncton Hospital. RPC performed a pre-construction review of mechanical plans for medical gas piping systems for compliance to standards and completed certification inspections upon completion of construction. RPC also conducts periodic inspections of medical gas systems as required to meet maintenance standards.

Another important support service provided to health care facilities is the provision of indoor air quality inspections and analysis. RPC regularly supports hospital maintenance and safety by providing inspection and analysis services for mould, asbestos and other indoor air quality concerns. Air quality is an important concern during construction and renovation. It is equally important to proactively 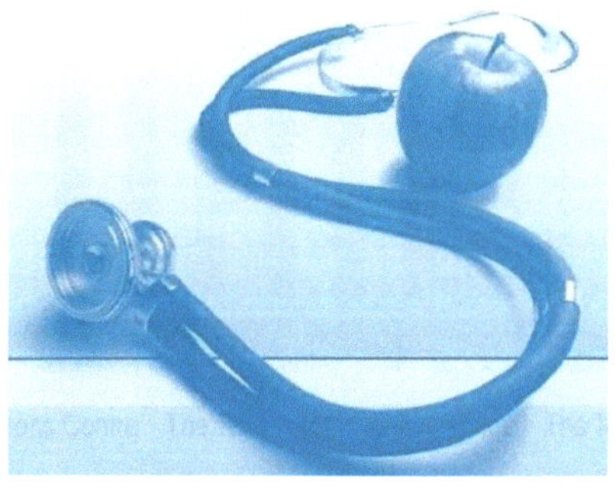 address health concerns through ongoing monitoring and maintenance. As with many of RPC's services, RPC is an important behind-the-scenes supporter of the safe operation of New Brunswick's health care facilities and has been helping to "keep the health in our health care system." See also "Health Matters," below.

Mineral Extraction Technology: RPC's 'Reign' in Spain (2006/07, [246]). The Aguas Teñidas Mine is located in southern Spain, 80 km WNW of Seville. Spain's northern pyrite belt, one of the world's most intensely mineralised areas, includes the Rio Tinto Mine with an estimated ore production of 500 million tonnes over a 2,500 year life. Discovering valuable ores can generate significant profits for companies and impact the economies of countries but only if economically viable and environmentally-compliant mineral extraction processes can be developed.

The developers of the Aguas Teñidas Mine engaged RPC to address the challenges associated with their ore resources. RPC is known throughout the mining industry for its strong capability with mineral extraction processes propelled by extensive experience with N.B. mines over the previous three decades. For this project, RPC employed flotation technology for the problematic polymetallic ore. Tonnes of the Spanish ore was shipped air freight to Fredericton for experiments which included optimizing of crushing, grinding and flotation, extensive analysis of the results, and dozens of iterations to perfect the process to produce an economically feasible grade and recovery rate. An added challenge was Spanish environmental standards that prevented the use of specified chemicals commonly employed in other countries.

Initial tests produced promising results and led to months of process optimization. Working closely with the client, diligent efforts employing extensive experience and sound science resulted in the discovery of an effective process and led to the development of a €200 million mine to be fully operational in 2008. The process incorporates very fine grinding and avoids the use of cyanide which has been banned from use in Spain due to

safety and environmental concerns. The advancement of RPC's flotation capability and the success with the challenges of this ore led to follow-on work and new clients, new equipment and new employees emphasizing, yet again, that New Brunswick technology is world class.

RPC: Energizing New Brunswick's Energy Hub (2006/07, [246]). The vision of a New Brunswick energy hub centered in Saint John was on the road to reality in 2006/07, with RPC as a significant contributor. Investments in the energy hub were forecast to possibly reach $15 billion. The Coleson Cove refurbishment ($750 million) was just finishing, while work was well underway on the electrical transmission line to the Maine border, the 145 km Brunswick Pipeline for natural gas, the liquefied natural gas (LNG) terminal, and the Point Lepreau power station refurbishment

($1.4 billion). Meanwhile, two other mega-projects were under consideration: a second oil refinery ($5-7 billion) and a second nuclear reactor, possibly the first installation of Advanced CANDU technology ($5 billion). The potential queue of such mega projects presented unprecedented business possibilities for N.B. industry and RPC's clients. New Brunswick's engineering firms, environmental consultants, construction companies, fabricators and manufacturers were all expected to benefit from the energy boom, with RPC positioned to continue to support them with critical technical services.

Examples of work RPC completed over 2006/07 include metallurgy analysis and testing for transmission line anchor bolts, weld procedure verification for the LNG terminal, precision flow measurements for the Point Lepreau nuclear generating station, feeder pipe crack inspections and pipe wall thickness investigations, environmental testing in support of the refinery site development, metallurgical inspections at Coleson Cove; and numerous analyses of emissions and by-products from various energy generation facilities. RPC was also actively engaged supporting next-generation nuclear technology, such as corrosion monitoring. Continuing research interests included on-line monitoring equipment that would permit more up time for energy facilities.

When Oil and Water Mix: RPC Helps Progress NB Innovation (2006/07, [246]). It is widely understood that oil and water don't mix, but when the two liquids coexist, effectively separating them can be a challenge. From his research conducted at Enterprise UNB, entrepreneur Dr. George Sutherland developed a potential solution employing a nanopolymer microencapsulating flocculating dispersion and hollow-fibre dehydration technologies to separate oil and water.

RPC worked with his company, Sutherland Separation Systems, to develop a bench-top demonstration of the technology and then test samples from several prospective application sites. The technology was successfully demonstrated and attracted significant interest. With the help of innovation funding agencies, a scaled-up field pilot was designed, constructed and tested at RPC. In early 2006, RPC worked with the client to conduct an important field trial of the technology at a British Petroleum site in Trinidad. That demonstration processed 20 litres per minute for approximately one month. The results were positive and demonstrated that the technology was effective at removing contaminants to a level well below legislated requirements.

Discussions were ongoing, in 2006/07, for commercial deployment of the technology and the success of the demonstration attracted interest from national and international clients.

Powerful Innovations (2007/08, [239]). By 2007/08, RPC had been delivering innovative engineered solutions and technical services to the nuclear industry for over 25 years. The successes continued in 2007/08, when RPC completed two projects that significantly advanced how non-destructive testing (NDT) is conducted for nuclear reactor feeder pipes.

The first such project involved a ***Magnetic Rubber Replica Kit.*** CANDU's impeccable safety and up-time records are achieved through excellent design, effective inspection, and maintenance. To aid in inspection, RPC developed a magnetic rubber replica kit for surface crack inspection in feeder pipes.

The project involved development of the inspection technique, design, construction and testing. The product utilizes clamp-on moulds with air bladders, a special miniature magnetic yoke and power box and a specialized rubber injection system. The magnetic rubber is injected and when removed, there is a physical replica of the crack which can be retained and utilized for future monitoring and comparison purposes.

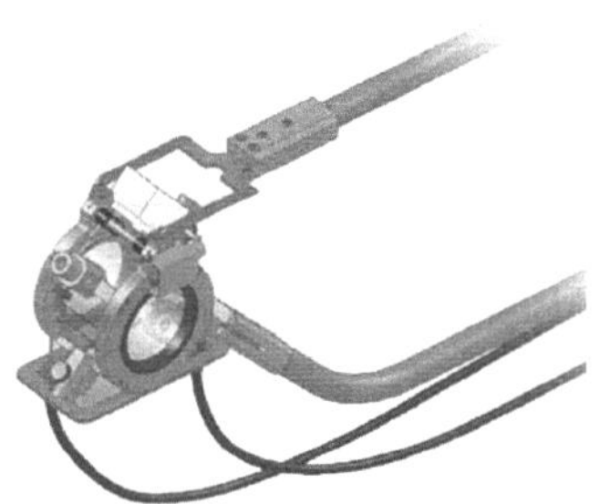

This tooling was developed for NB Power and Hydro Quebec and became the accepted CANDU Operator's Group method to detect and archive tiny surface cracks on feeder pipe bends.

The second innovation was the development of ***Remote Magnetic Particle and Radiographic Inspection Tools*** for AECL. These were developed to complete single feeder replacements in very difficult to access areas of a nuclear reactor. This project involved the complete design and construction of 12 sets of tooling to the rigorous Z299.2 quality assurance standard. The tools have been employed at Bruce Power, Pickering and Darlington nuclear plants in Ontario. The reviews from the field were very good and, in 2007/08, RPC was developing concepts for next-generation tooling for the new CANDU ACR 1000 reactor.

Health Matters: RPC Medical Gas Certification Service Helps Atlantic Health Care Facilities (2008/09, [249]). As already noted above, one of the many technical services offered by RPC is accredited medical gas certification. When hospitals and health care facilities complete new construction or renovations that involve medical gas piping, they are required to have a medical gas inspection completed by a qualified provider prior to commissioning the system. RPC's medical gas inspection staff are highly experienced in all aspects of medical gas piping systems, from concept design to completion and testing. RPC is the only accredited service provider east of Ontario, and is recognized by the Standards Council of Canada for the inspection of non-flammable medical gas piping systems and the analysis of compressed breathing air.

RPC staff work diligently throughout Atlantic Canada to provide professional and timely service to facilitate the improvement of the region's health care facilities. Medical gas certification is one of the many services provided by RPC that otherwise would have to be obtained from outside N.B. See also "Keeping the Health in Our Health Care System," above.

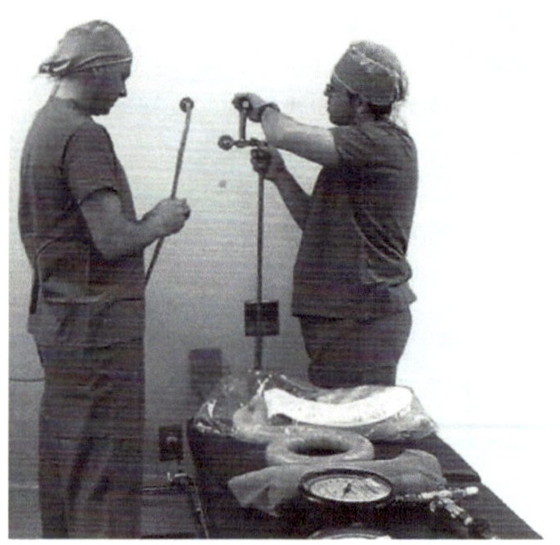

CSI Deep Woods New Brunswick: RPC Launches Forensics Biology Lab (2009/10, [255]). In April 2009, RPC launched its Forensic Biology Lab, with Standards Council of Canada (SCC) accreditations for wildlife forensics and relationship analysis. Working with the Department of Natural Resources, RPC became the first lab in Canada to be SCC accredited and is thought to be the first in North America to gain SCC 17025 compliance. The forensic services for wildlife include capabilities for species identification, DNA fingerprinting and gender typing with direct applications to assist wildlife enforcement and protection, and wildlife researchers. In New Brunswick, wildlife forensics facilitate the prosecution of poachers, helping to protect moose and deer populations.

Over the 20 years prior to 2009/10, RPC developed DNA capabilities in support of New Brunswick's aquaculture sector and established world-class services helping industry to flourish. Building on this expertise and capability, RPC worked with subject matter experts to develop wildlife forensic capabilities, and more recently, human paternity testing.

DNA science has been made famous through television crime shows, although the shows oversimplify a process that actually requires a rigid quality assurance process, comprehensive chain of custody procedures for handling samples, and rigorous management of data to protect confidentiality. See also "Whose Moose is Loose?" above.

The Sound of Success: RPC's Ultrasonic Technology (2010/11, [254]). By 2010/11, RPC had been developing and applying its expertise in ultrasonic flow metering for more than two decades. Initially developed for application in the nuclear industry, RPC's flow metering expertise has been applied in New Brunswick's food processing industry, pulp and paper industry and even a geo-thermal heating system. During 2010/11, a significant success was achieved when RPC was awarded a three-year service contract to design, manufacture and install a high accuracy (>99%) flow and temperature meter at Bruce Power in Ontario. This application required a comprehensive quality assurance regimen and reliability in a harsh environment that includes operating temperatures reaching 260 °C. RPC's innovative use of ultrasonic technology included temperature measurements co-located with flow measurements. The project required the design and construction of innovative fixtures and high temperature transducers, as well

as the development of control software and measurement procedures. The calibration process demonstrated accuracy in excess of requirements and the systems were installed to take measurements beginning in March 2011. RPC's success in providing an innovative solution to meet an industry need then generated additional inquiries for precision temperature and flow measurement.

'Bacteria' to the Future (2007/08, [239]) and **Bugs as Drugs: Probiotics Have Human Health Potential** (2011/12, [261]). Probiotics are bacteria that confer benefit to their environment. Already widely employed in food products such as yogurts, probiotics have many other applications including mineral extraction, environmental processes and aquaculture. RPC led a project to develop probiotics and other novel bacterially derived products for use in aquaculture.

This research, initiated by an industry need and funded by ACOA, has allowed RPC to establish significant expertise in the areas of probiotic discovery, testing, and application, and in the related area of bio-prospecting. By 2007/08, RPC had established a collection of over 1,500 bacteria derived from a variety of coastal N.B. marine environments. By 2011/12, the collection had grown to over 1,800 bacteria. Many of the bacteria in the library appear to be new species as well as those that have not been previously cultured. These bacteria have been screened for probiotics and antimicrobials for use in aquaculture. With such a relatively large number of bacteria, RPC has developed a high-throughput screening methodology, enabling the entire library to be screened using a variety of conditions. This has greatly increased the probability of discovering candidate strains for further study.

A promising application is for improved fish larval survival rates. In Atlantic Canada and elsewhere, rearing of alternative species such as cod and haddock has often been plagued by very low (3-5%) survival rates at the larval stages. The use of probiotics developed within this project has resulted in significant improvements in larval survival. This improvement is achieved through the ability of the probiotics to encourage the development of a healthy bacterial community with the larvae and exclude the establishment of harmful pathogens. This green solution to a common hurdle in the rearing of alternative species is expected to boost the productivity of the aquaculture industry in Atlantic Canada and provide a viable alternative to the use of antibiotics.

In more recent work, a collaboration with MARBIONC Development Group in North Carolina, identified potential for human health applications such as cancer treatments and new antibiotics. RPC announced this discovery in January 2012, leading to national radio, television, and newspaper stories and interviews. RPC's Dr. Ben Forward observed that, *"The neat thing about bacteria is that they can have quite complex biochemistries, which allows them to synthesize some very complicated molecules that you probably couldn't synthesize in a lab."* Land-based microbials have already led to important drug discoveries. The ocean represents a new and expansive source of new biomolecules that may lead to a next generation of human health products.

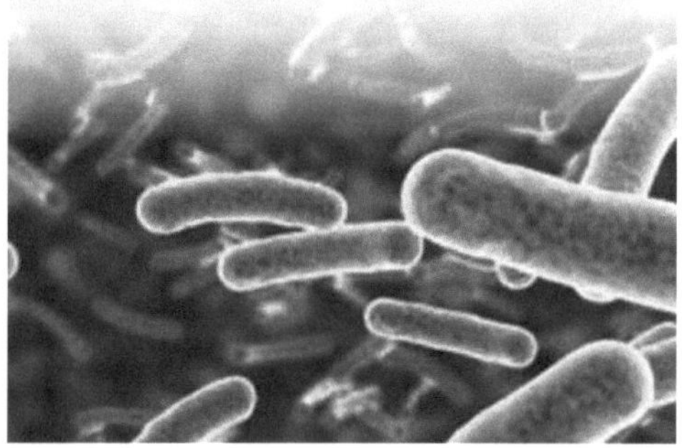

Hot Topics: RPC's Comprehensive Analytical Chemistry Services (2013/14, [263]). RPC has long maintained an extensive scope of accreditation for analytical services including a comprehensive suite of inorganic and organic analytical chemistry. Everyday analyses are performed to support business, industry and all three levels of government. Sometimes, a topic is 'newsworthy,' meaning that it is fairly current, affects the community, and often occurs 'close to home.' When a topic is newsworthy, RPC's analytical services are often involved in delivering associated services. Some examples from 2013/14 included ***medical cannabis***, for which RPC is one of the very few commercial labs in the country offering accredited testing for the full suite of required parameters (including the organic, inorganic and microbiological analyses), ***shale gas***, for which methane in water is a frequent concern and for which RPC is fully accredited for testing (plus related organic and inorganic analyses), ***radon gas***, a leading cause of lung cancer and for which RPC is certified to conduct the analyses, ***N-Nitrosodimethylamine (NDMA)***, a health concern in some water and wastewater supplies and for which RPC is accredited to conduct the analyses, and ***polybrominated diphenyl ethers (PBDEs)***, formerly a component of flame-retardants but still a health and environment concern and for which RPC is accredited to conduct the analyses. As the examples show, if a hot topic appears in the news and presents a public concern, there is a good chance that RPC is already involved in providing accredited analyses.

GENETIC IDENTITY: Monitoring Microbial Communities with eDNA for Environmental Management and Ecosystem Health (2017/18, [265]). Environmental DNA or eDNA is DNA that is collected and extracted from holistic environmental samples rather than directly from individual organisms. The DNA may come from the individual organisms contained within these samples, or it may be free-floating DNA that has been shed by organisms in the area. Sampling and analyses can determine the environmental biodiversity within a sample and ongoing monitoring can be employed to establish a baseline for detection of changes to that environment over space and time. RPC has been working with Department and Fisheries and Oceans (DFO) scientists to study bacterial biodiversity in the marine environment in the Bay of Fundy in relation to anthropogenic stressors on ecosystems. This approach uses an innovative sampling device based on remote operated vehicle (ROV) technology developed by DFO to sample the surface of the sea bottom together with RPC's next generation sequencing capabilities for bacterial identification.

Results have been very promising and are consistent with recent results from other initiatives going on internationally. Ultimately, this could become a broad-based tool that can be widely used for environmental monitoring in a variety of habitats. This is an excellent example of an eDNA application made possible with next generation biotech analytics and creative approaches to large-scale questions.

(Photo Credit: S. Robinson Fisheries and Oceans Canada)

6.4 RPC's Long-Standing Clients.

One of the stories that does not fit neatly into any single one of the denoted eras of RPC's evolution, is that of its long-standing client relationships. Several of these relationships have transcended these eras of evolution and remain healthy to the present day. Three top-of-mind examples are represented by McCain Foods, Cooke Aquaculture, and NB Power.

McCain Foods Ltd. Prominent on RPC's first-ever list[60] of industrial clients, reported for 1965/66 [55], is McCain Foods Ltd., and in fact RPC has had an active relationship with McCain Foods since inception in 1962. Not only did the company become one of RPC's first clients, its president, Harrison McCain, was a founding Board member of RPC.

The first projects for McCain ranged from potato-storage technology to colour-control in potato chips [290]. Subsequent programs included the development of new potato-based processed-food products (Figure 6.2), new French Fry designs (Figure 6.3), automated cutting machines (Figure 6.4) and many others.

More recently, in 2014/15, RPC partnered with McCain on a planned $10 million[61], five-year project to engage with SMEs and other N.B. researchers to pursue process improvement, product development and soil remediation initiatives [256]. Examples included ways to reduce waste through technological means, and soil remediation and agrological analytics using drone technology.

---

[60] Of the fifteen contract-service clients in 1965/66, nine were from industry and six were government organizations.

[61] The provincial government invested $5 million, with McCain providing $5 million in matching funds.

Figure 6.2. Laboratory testing in support of a McCain processed-food product, in this case puréed potatoes. RPC photo, 1986 [189].

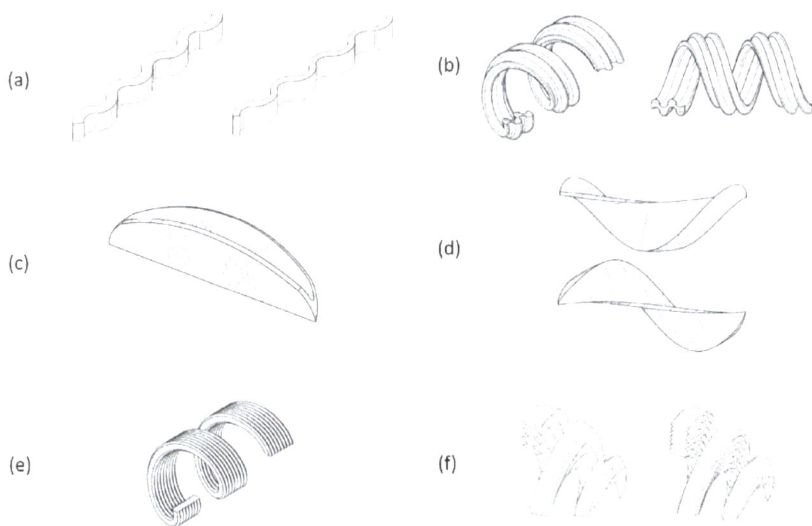

Figure 6.3. Illustrations of some of the French Fry shapes designed for McCain Foods. These can be cut by the kind of automated cutting machine developed by RPC (see Figure 6.4). The shape designs are from U.S. Design patents: (a) D707,418, 2014; (b) D712,618, 2014; (c) D714,521, 2014; (d) D720,916, 2015; (e) D722,417, 2015; (f) D748,366, 2016 (see Appendix 8.8).

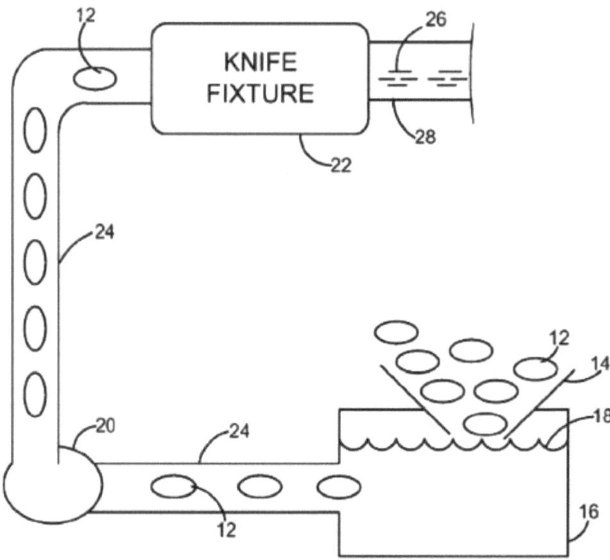

Figure 6.4. Schematic drawing of an improved hydraulic French Fry cutting system developed for McCain Foods. Here, potatoes (12) are fed from a hopper (14) into a tank (16), in which they are submersed in water (18). Multiple pipes (24) connect the tank to a pump (20), and from there to a knife fixture (22) at high velocity. The knife fixture includes a blade assembly designed to cut the potatoes into smaller pieces of specified shape (26) and discharged through an outlet conduit (28). From U.S. Patent Application 15/454,552, filed April 1, 2014. The complete application citation is given in Appendix 8.8.

Cooke Aquaculture. Founded in St. George, N.B. in 1985, Cooke Aquaculture Inc. grew to become a fully-integrated producer of Atlantic Salmon products. In 2004, it billed itself as North America's largest independent salmon farming company and *"experts from egg production to recipe formulations"* [291].

The aquaculture sector, including Cooke Aquaculture and its subsidiaries, embraced science and innovation from its inception. RPC's Dr. Bev Bacon recognized the opportunity and helped RPC become an early leader in the science, technology, and innovation side of the business. Cooke was not only an early client from this sector, but has remained so to the present day.

In March 2008, when interviewed for a N.B. business publication Nell Hulse, Cooke's Vice President of Communications declared: "*Thankfully, [in RPC] New Brunswick has one of the greatest R&D facilities in the world supporting the private sector*" ([292], also cited in [239]).

In one of the more recent projects, RPC has been helping Cooke Aquaculture to develop an innovative DNA-based traceability system to track individual Atlantic salmon from 'farm-to-fork.' Fish traceability is important given growing public concerns about food safety, new regulatory requirements and increasingly competitive markets. The new system will incorporate DNA technology to facilitate tracking of Cooke's stock and individual fish within the hatchery and throughout all phases of production, and will determine the unique 'DNA barcode' for each fish, enabling its identification and traceability throughout its life. As a first step, RPC was able to identify unique areas of DNA that are common to the offspring of the broodstock, enabling the tracing of fish back to the parents by batch.

NB Power (New Brunswick Power Corp.). Founded in 1920, NB Power needed RPC's help when it expanded its electric power production beyond coal and oil to nuclear in the 1970s. Nuclear power plant construction at its Point Lepreau site started in May 1975 and was completed in 1981 [293]. RPC was asked to assist from the very beginning, and established a non-destructive testing and physical metallurgy group in 1975. As described in Chapter 3, RPC staff began working on non-destructive evaluation and metallurgy, and ultrasonic flow metering for NB Power in 1976.

As described in Chapter 4, such work continued through the 1970s and 1980s, culminating in the establishment of the Centre for Nuclear Energy Research (CNER) – a partnership of NB Power, RPC, and UNB – in 1992. With CNER, RPC was able to continue to work with NB Power on a broader range of problems related to the safety and operation of its nuclear generating plant. Although CNER was reorganized as a research institute within UNB in December 2010, RPC has continued to be engaged as a collaborator within CNER and also as a service provider to NB Power.

Some specific, modern-day stories illustrating RPC's work for NB Power have been described above in Section 6.3 (Figure 6.5).

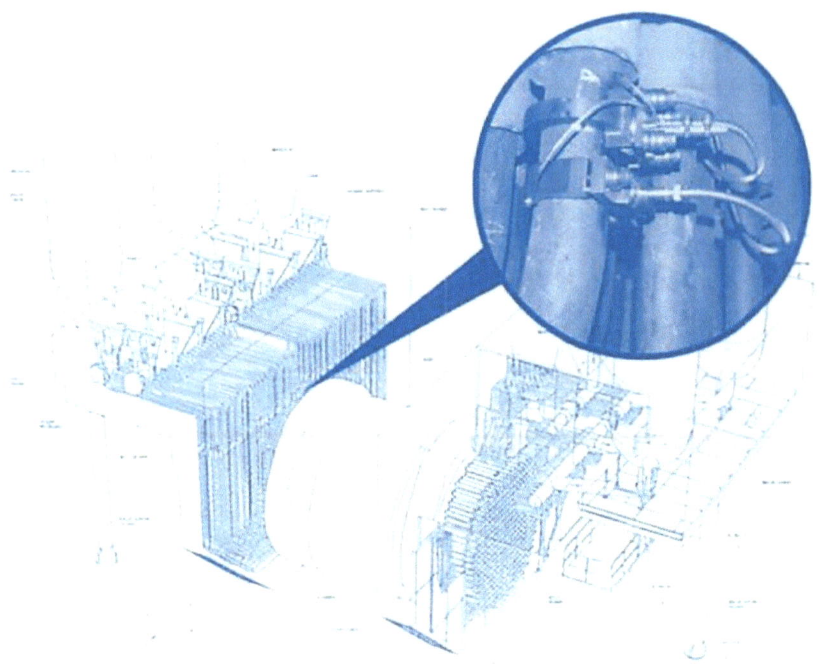

Figure 6.5. Illustration of a corrosion monitoring system for NB Power's Point Lepreau CANDU nuclear generating station that was developed, tested, and manufactured by RPC (*circa* 2006 [245]).

6.5. Corporate Social Responsibility.

All of RPC's work is aimed at benefitting the people and communities of New Brunswick, mostly by helping businesses in the province to strengthen the economy, with quality jobs. There is, however, another side to RPC's service to the public-good: that of community service.

Although *ad hoc* community volunteer efforts on the part of RPC employees probably date back to RPC's inception, larger-scale, corporately-supported Corporate Social Responsibility (CSR) activities have been undertaken in *The Entrepreneurial Years* to complement the many individual charitable efforts led by RPC employees.

When people think of CSR, they often first think about charitable community initiatives. The United Way, supported by RPC employee contributions, has been RPC's primary corporate charity followed closely by the Fredericton Food Bank (Figure 6.6). Examples of other community support have included helping the Dr. Everett Chalmer's Hospital, Big Brothers and Big Sisters, Opal Family Services, Camp Rotary, women's and men's shelters, Red Cross, Canadian Blood Services, SPCA, and Help for

Haiti [246,255].

Figure 6.6. Examples of community initiatives supported by RPC employee contributions.

Other kinds of CSR initiatives include supporting and assisting RPC employees to get involved with:

o Participating on governance or advisory Boards in the community. In 2010/11, for example, these included the Atlantic Canada Fish Farmers Association, Genome Atlantic, the Center for Nuclear Energy Research, and the NB Environmental Industries Association [254],

o Participating in trade and professional associations, standards committees, and university and college committees. In 2010/11, for example, these included a breathing air standards committee, a medical gas standards committee, a N.B. Community College curriculum committee, and a Canadian Council of Academies committee [254],

o Supporting the annual Take Your Kids to Work Day, Junior Achievement and, naturally, serving as judges in science fairs.

While it will be noted that some of these activities also bring benefits to RPC and the individual employees concerned, they also have a component of charitable community support and, in the case of the individual employees, typically comprise substantial components of individual (unpaid) volunteer time.

RPC's first formal approach to CSR began with its Strategic Plan 2015-2020, within which CSR was identified as a strategic objective [256]. In addition to its primary goals of supporting the United Way and Fredericton Food Banks, RPC would continue to encourage and support employee-led charity initiatives, and expand its environmental footprint reduction (e.g., reducing solvent use energy consumption) and recycling initiatives (from paper to batteries to pipette tips).

UNITED EFFORT: RPC Community Giving (2016/17, [273]). RPC employees give back to the community in many ways. Some are members of service clubs, others are involved in school and church initiatives, others volunteer directly with charities and community initiatives. RPC's official corporate charity is the United Way. In 2016/17, RPC saluted Karen Broad for her service as Team Captain of RPC's annual United Way Campaign and thanked RPC employees for their participation. Karen, working with a team of volunteers, coordinated RPC's activities, communicated the goal, distributed pledge forms, and encouraged donors. RPC also noted that it was *"especially proud of the participation of our new employees, many who are starting their careers and have become new donors, helping RPC to achieve its increasing goal."*

Figure 6.7. Karen Broad, RPC's United Way Team Captain in 2016/17.

7 RETROSPECTIVE

The Past. RPC has come a long way since its inception in 1962. Created to "stimulate and expedite continuing improvements in productive efficiency and expansion in various sectors of New Brunswick's economy," RPC would use research, development, and technological services to work with industry to help grow and diversify the New Brunswick economy. The manner in which it did so, however, had to change in concert with changing environments and changing industry needs.

At first, in its *Building Years* (1962-1969), RPC got much of its work accomplished through a 'Research Program,' by which grants were awarded to UNB faculty for specific projects aimed at the province's diverse natural resources, including fisheries, and manufacturing, and through a 'Productivity Program,' that involved training personnel from industry in modern management and industrial production techniques. It soon became necessary, however, for RPC to provide applied R&D, analytical and testing services that were beyond the capabilities of the university, driving RPC to establish its own dedicated scientists, engineers, and technologists, specialized instruments and machines, and the necessary laboratory and pilot plant infrastructure in which to conduct the work. A critical mass of such people, tools, and infrastructure was in place by 1969 and, having conducted hundreds of projects for industry, RPC's ability to serve N.B. industry was becoming widely recognized.

With a solid base of staff, tools, and infrastructure, and annual revenues in the range of $800,000, RPC entered its *Growing Years* (1970-1983) well positioned to show what it could do. In this era, RPC built up a diversified portfolio of services that ranged from technical information services and training, to analysis and testing, to feasibility studies, development engineering, industrial engineering, and commercialization. In this era, in particular, RPC built up capabilities – and a solid reputation – in pilot- and demonstration-scale plant design, construction, and operation, enabling it to take on larger projects for major industry and even nucleate and spin-out companies of its own. By the end of this era, in 1983, RPC had matured as an organization, its staff level had essentially doubled, and revenues had grown five-fold, to nearly $5 million per year.

A change of strategy altered its course somewhat in its *Commercial Years* (1983-2004) era. As the name implies, this era involved learning how to operate more like a business and derive increased revenue from contract services. As revenues continued to grow, additional laboratory and other facilities were added, as were additional technical capabilities, especially in metallurgy, manufacturing technologies, aquaculture, and biotechnology. RPC's strength and positioning enabled it to play an important role in three public crises of the era: the 'Salmon Crisis,' the 'Rancid Tuna Scandal,' and the 'Spruce Budworm Crisis,' which also served to enhance its reputation with all stakeholders. Also noteworthy in this era was the special focus that was placed on helping N.B. entrepreneurs and SMEs. By about 1997, RPC was serving 300 or more SMEs per year, a level from which it has probably never retreated. This brought balance to RPC's long-standing work with large companies, and with the next wave in RPC's evolution: extra-provincial work. With its clients increasingly heeding the trend towards globalization, RPC also increased the levels of work it undertook for companies across Canada and around the world (Figure 7.1). This was a healthy expansion for RPC as it not only increased and diversified its revenue base, it helped make RPC world-class in the services it offered to N.B. clients. By the end of this era, RPC had almost completely transitioned to contract services, and RPC's own productivity had dramatically increased: from a stable staffing level, compared with the previous era, revenue doubled over this period. RPC had become diversified and efficient.

RPC's *Entrepreneurial Years* (2004-2020) began with a strategy that focused on industry more broadly than the previous focus on SMEs. A recognition that globalization, shortened product life cycles, and rapidly changing economic and political environments meant that RPC, and the businesses it intended to serve, would have to be increasingly prepared for change. Accordingly, RPC focused on becoming more agile: anticipating and responding to change, offering ever more diverse business services, emphasizing collaborative efforts where possible, and seeking to manage the risks inherent in attempting to work on the leading-edge of technology without slipping over the 'bleeding edge.' These changes helped the organization withstand the next three major challenges that would appear: increased competition from out-of-province laboratories, increased competition from within the provincial government itself, and the decline and, ultimately, elimination of core operational funding support from the province.

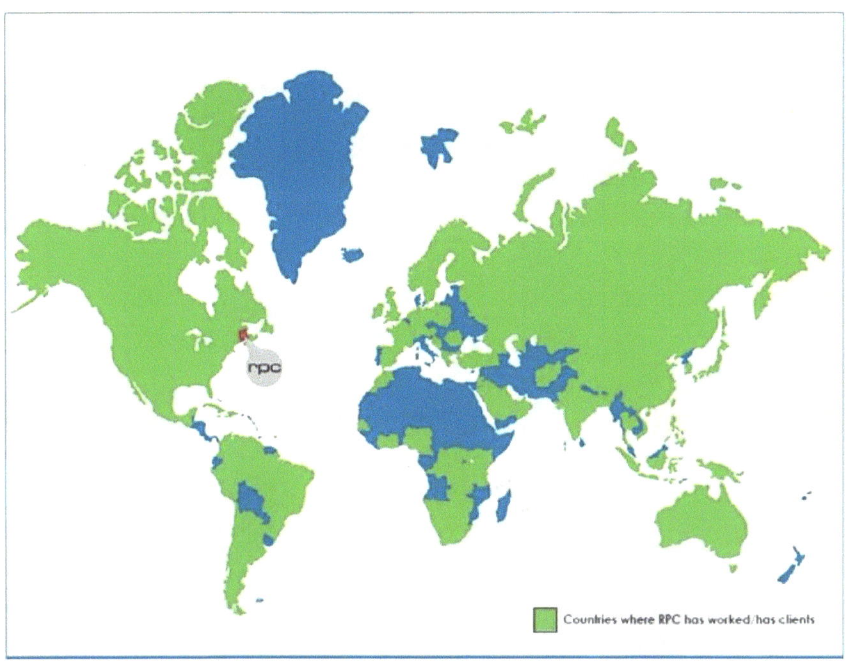

Figure 7.1. Illustration of RPC's global reach. The areas highlighted in green represent the countries in which RPC has, or has had, clients. International work continues to come in to RPC from about 30 countries each year.

At the same time, new and improved services were developed and offered to the marketplace in such areas as aquaculture, air quality testing, environmental-chemical analyses, and genetics. To these were added a number of major expansions in capability in order to position RPC to assist manufacturers' transition to 'Industry 4.0,' including machine learning, machine vision, and robotics. As it turned out, the changes adopted in RPC's *Entrepreneurial Years* not only enabled it to withstand the three organization-threatening changes, they helped it to grow. By 2019, RPC's capabilities and capacity were at an all-time high, annual revenues had doubled to over $16 million (compared with the previous era), and the staff level had grown by 70% to approximately 160.

The one constant through all four eras has been RPC's ability to respond to N.B. industry's needs, deliver practical solutions at the speed of business, and achieve positive impacts in New Brunswick's economy. Chapter 6 presents a selection of such impacts and some of the stories behind them.

In carrying out its mandate, RPC has helped industry develop and advance, supplying everything from analysis and testing, to expert assistance with the full product/process development cycle – from concept to commercialization. Although not among the first of Canada's RTOs to be established, RPC has evolved to become one of the most enduring, and one of the nation's leaders in the modern approach to technological innovation.

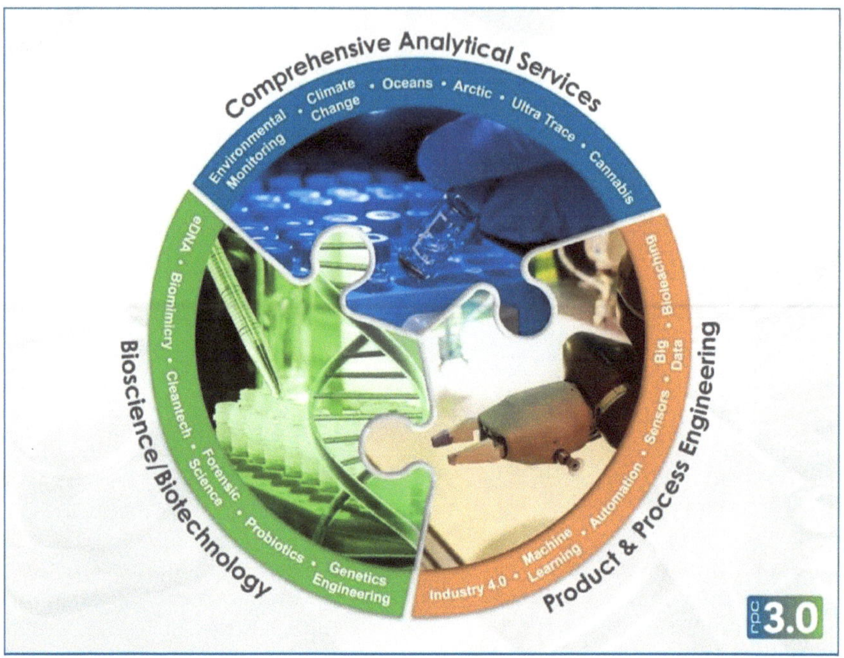

Figure 7.2. Illustration of the range of service lines provided by RPC in response to requests from clients in New Brunswick and around the world.

The Future: RPC 3.0. RPC's *Annual Report 2017/18* noted [265] that: *"Enormous workloads and rapid growth have accelerated the end of life of key infrastructure. Planning is underway for RPC 3.0, a project to revitalize, enhance, and expand [RPC's infrastructure]."* This work, coupled with efforts to secure the necessary funding were still underway at the time of RPC's Annual Report 2018/19 [280] and, indeed at the time of writing. From an infrastructure perspective:

o RPC 1.0 refers to the first six buildings that were built or acquired during *The Building Years* (1962 – 1969), and up until about 1987 in *The Commercial Years* (1983 - 2004),

o RPC 2.0 refers to the addition of Building #7 in 1991 plus the acquisitions of the Moncton (2010) and St. George (2017) laboratory buildings during *The Entrepreneurial Years* (2004 - 2020), and o RPC 3.0 refers to the currently planned, major infrastructure upgrade and/or expansion.

RPC 3.0 contemplates realizing the infrastructure needed to support the market demands for RPC services in Bioscience/Biotechnology, Comprehensive Analytical Services, and Product & Process Engineering (Figure 7.2).

In this vein, it seems appropriate to give the final words to New Brunswick Premier Louis J. Robichaud who, in 1965 [1], said:

"An important part of the task of [RPC]… is to exercise careful judgement in choosing fields of research which have a good probability of increasing productivity in existing or new industry… [and that] the government expects assistance to New Brunswick industry in the application of available technology, and the development of new techniques of direct application in the province. Such service will, without doubt, contribute significantly to increased production and productivity and a better standard of living for us all. Expansion, change, new developments – these are becoming the order of the day in New Brunswick."

…and also, to RPC's first Chair, Dr. Les Shemilt, who, in 1968 [57], said:

"Growth is the only evidence of life. RPC does and will have life. We must and will continue to grow…"

validation
excellence
competence
documentation
confidence
certification
audit
ISO
continuous improvement
training
quality
method
calibration
records
standard
SOP
proficiency test
assessment
assurance
repeatability
control chart
process
safety

8 APPENDICES

Appendix 8.1. An RPC Chronology.

1962 – 1969, *The Building Years*
1970 – 1983, *The Growing Years*
1983 – 2004, *The Commercial Years*
2004 – 2020, *The Entrepreneurial Years*

Year(s)	
1916	First broad, multi-sectoral research and technology organizations in North America, the U.S. NRC, and the Canadian NRC were launched.
1962 - 1969	*The Building Years* at RPC
1962	Creation of RPC through an Act of the New Brunswick Legislature.
1963	First full-time Executive Director and three other full-time employees hired.
1963	Technical Information Service (TIS) for industry program delivery in New Brunswick initiated.
1964	TIS program delivery extended to include Prince Edward Island.
1964	RPC joint-ventures with UNB and other agencies to form the Atlantic Provinces Institute of Forestry, to encourage higher productivity of utilization of forest resources.
1965	First permanent laboratory building opened ('Building #1').
1966/67	First high-head pilot-plant building opened ('Building #2'); Additional laboratory block acquired ('Building #3').
1968	Additional block of laboratories and pilot-plant space built ('Building #4').
1970 - 1983	*The Growing Years* at RPC
1973	Innovent Ltd. subsidiary created to stimulate pilot-plant activities.
1973	Sulphation-roast-leach (SRL) electrowinning technology for the processing of low-grade copper-lead-zinc ores developed.
1974	Galleon Ware venture launched, involving the development and manufacture of ceramic tableware. Made a subsidiary in 1975.
1975	Minuvar Ltd. subsidiary created to enable new mineral resource development and (later) soil remediation.
1975	Opening of 20,000 sq. ft. Production Transfer Pilot Plant (PTPP), 'Building #5.'
1975	Non-destructive testing and physical metallurgy group established to assist Point Lepreau nuclear power plant.
1978	Royal Coat of Arms and Badge granted by Letters Patent.

Year(s)	
1979	Mobile energy audit program (the 'energy-bus' program) launched in N.B. to help companies reduce energy consumption.
1982-1986	Coal/oil shale work, involving fluidized-bed combustion, leads to $35 million Circulating Fluidized Bed demonstration plant at Chatham, N.B. by the N.B. Electric Power Commission.
1983	Creation of ERS and larger plant for RPC's SRL process.
1983	$2.5 million project to develop a prototype rechargeable lithium battery.
1983	Enhanced Recovery Systems (ERS) subsidiary created to advance pilot-plant-scale development and demonstration of mineral processing technologies, including the SRL process.
1983 - 2004	*The Commercial Years* at RPC
1984	Demonstration centre for CAD/CAM launched
1985	RPC assists government and industry in *'Tunagate'* as an independent, qualified 'third party,' safety-testing canned tuna, and helping industry improve product-inspection procedures.
1985	RPC joint-ventures with UNB, Université de Moncton, and three colleges to form the Manufacturing Technology Centre (MTC), to transfer CAD/CAM technology to N.B. companies.
1985-1986	Decentralization, with RPC opening regional offices in Saint John, Moncton, Newcastle, and Edmundston
1986-1987	Development of a hermetically-sealed, rechargeable 'C-cell' prototype, ready for pilot-plant production, commercialization.
1988	8,000 sq. ft. administrative office wing ('Building #6') opened.
1988	Major metallurgical program undertaken for the Department of National Defense's Canadian Patrol Frigate Program spanning welding, material selection/inspection, and RCN officer training.
1988	RPC joint-ventures with UNB to form Incutech Brunswick Inc., to support entrepreneur-led technology-intensive manufacturing businesses in N.B.
1989-1990	With continuing decentralization, regional offices were opened in Shippagan and Bathurst.
1991	18,000 sq. ft. state-of-the-art laboratory building ('Building #7') opened.
1991	*Agglutitest* diagnostic kit, developed for the rapid, inexpensive detection of diseases such as Bacterial Kidney Disease in salmon licensed to Valox Ltd. for commercial production.
1992	RPC joint-ventures with UNB and NB Power to conduct R&D related to the safety and operation of nuclear generating plants.
1994	A field test of RPC's synthetic spruce-budmoth sex-pheromones at a J.D. Irving Woodlands white spruce plantation at Boston Brook, N.B., achieves 80% reduction in viable egg production.

Year(s)	
1997	RPC assists government and industry in *'The Salmon Crisis'* by testing fish and by identifying the virus that was killing Bay of Fundy salmon.
2001	Company-wide ISO 9001:2000 registration obtained.
2001	RPC scientists discover a naturally-occurring bacteria that helps fish develop an immune response to Bacterial Kidney Disease, and licenses the vaccine product, Renogen®, for sale worldwide.
2004 - 2020	The Entrepreneurial Years
2004	RPC develops probiotic bacteria for use in hatchery culture of marine finfish and shellfish, alleviating a critical production-bottleneck while improving productivity and sustainability.
2005/06	With increased global reach, RPC's clients span 33 countries beyond Canada.
2008	To advance DNA-based traceability for farmed salmon, RPC begins test-tracking batches of fish from 'farm-to-fork.'
2009	RPC becomes first in N.B. to offer accredited human-DNA-parentage-testing services.
2011	Acquired environmental analytical laboratories in Moncton from Caduceon Environmental Laboratories.
2014	Achieved licensing and accreditation (and a leadership position) for analyses required by medical cannabis producers.
2016-2017	Acquired provincial dairy, fish health, and environment labs.
2016-2017	Acquired toxicology lab from Buchanan Environmental Ltd.
2018	RPC creates a Chief Science Officer position, joining a growing trend across Canada.
2018/19	Advanced manufacturing technology adoption program launched to identify company-specific opportunities for automation, robotics, machine vision, data collection, process automation, data analytics, 3D-printing, machine learning and other advanced, 'Industry 4.0,' manufacturing technologies.

Appendix 8.2. Members of the RPC Council/Board of Directors[62].

Member	Years Served
Addison, Mr. J.	1970 – 1975
Andrew, M. J.T.G.	1981 – 1988
Armstrong, Mr. L.S.	1983 – 1989
Arnold, Mr. W.	1991 – 1994
Arsenault, Mr. F.J.	1981 – 1984
Arsenault, Mr. W.	2014 – 2017
Aucoin, M. F.	1991 – 1995
Bacon, Dr. B.	2008 – 2014
Beattie, D.	2008 – 2014
Beatty, S.	2008 – 2014
Bernier, M.	2004 – 2006
Blair, Mr. D.W.	1981 – 1987
Bliss, Mr. T.	1995 – 1999
Blanchard, Mr. J.P.	1975 – 1983
Boorman, Dr. R.S.	*ex officio* (Executive Director) 1983 – 1992
Borland, W.	2003 – 2008
Bouchard, Mr. G.	1998 – 1999
Boucher, Dr. N.	2004 – 2010
Boudreau, Jean	2003 – 2008, 2014 – 2017
Brennan, Mrs. A.	1988 – 1991
Bujold, Mr. S.	1972 – 1975
Burchill, Mr. J.	1964 – 1976
Burns, Dr. D.	2014 – 2017
Burridge, Dr. R.	1976 – 1991
Bursill, Dr. C.	*ex officio* (Executive Director) 1963 – 1983
Caissie, Mr. N.	1995 – 1999
Campbell, Mr. D.C.	1973 – 1975
Caron, D.	2014 – 2015
Cassidy, Dr. S.	1976 – 1988
Carter, Mr. G.	1988 – 1991
Celeste, Mr. L.	1991 – 2003
Chiasson, M. V.	1988 – 1995
Comeau, A.	2016 – Present

---

[62] Originally, the Board was referred to as the 'Council,' and the Members were 'Members of Council.' Beginning with the 1989/90 fiscal year the terminology was changed to 'Board of Directors' and 'Directors', respectively.

Member (cont'd)	Years Served (cont'd)
Cook, Mr. E.	*ex officio* (Executive Director) 2004 – 2017[63]
Corey, Mr. L.	2010 – 2016, Chairman 2015 – 2016
Cormier, Mr. M.	1995 – 1998
Cox, Mr. K.V.	1962 – 1991, Chairman 1970 – 1986
Cragg, Dr. L.H.	1963 – 1969
Craig, Mr. J.	1991 – 1995
Crawford, B.	2008 – 2014
Day, Mr. J.	1991 – 2001
Desjardins, Dr. L.	1992 – 1998
Dick, Mr. B.	1991 – 2008
Doucet, Ms. L.	1995 – 1998
Doucet, Mr. M.	1973 – 1976
Drummie, Mr. F.R.	Executive Director 1962 – 1963; Member 1962 – 1966
Dubé, J.	2003 – 2004
Duguay, É.	2007 – 2008
Estes, Mr. W.E.	1962 – 1967
Ettinger, D.	2008 – 2014
Eusanio, Mr. R.	1995 – 1999
Finn, J.	2016 – 2017
Finn, Mr. G.	1969 – 1971
Flett, Mr. L.W.	1962 – 1964
Fredericks, Mr. H.	1962 – 1963
Gagnon, J.	2014 – Present
Gallagher, Mr. D.W.	1962 – 1963
Grotterod, Dr. K.	1985 – 2008, Chairman 1986 – 2008
Hargrove, Dr. L.	2016 – Present
Hesler, Dr. N.	1962 – 1963
Jaeger, Dr. L.G.	1972 – 1975
Jones, W.M.	2003 – 2008
Jennings, J.	2014 – Present
Kealey, Dr. G.	2001 – 2008, 2010 – 2013
Kenny, Mr. M.	1962 – 1973
Lajoie, Mr. L.H.	1962 – 1965
Lanteigne, Rev. L.	1962 – 1969
Laplante, Mr. D.	2010 – 2016
LaPointe, K.	2003 – 2005

---

[63] The CEO/Executive Director Position has not been an *ex officio* board position since the RPC Act was updated in 2017.

Member (cont'd)	**Years Served** (cont'd)
Larochelle, C.	2018 – Present
LeBlanc, B.F.	2008 – 2011
LeBlanc, Ms. J.	1988 – 1991
LeBlanc, Dr. L.	1976 – 1991
Lepage, P.	2008 – 2009
Levesque, Mr. B.	2010 – 2014
Lewell, Dr. P.	*ex officio* (Executive Director) 1992 – 2004
Likely, Mr. J.A.	1976 – 1982
Long, Dr. Y.	1965 – 1969
MacBain, Mr. I.	1972 – 1975
MacFarlane, T.	2019 – Present
McCain, Mr. H.	1962 – 1968
McCardle, I.	2016 – Present
McGladdery, Dr. S.	2010 – 2014
McGuire, Mr. F.	1991 – 1998
McIntosh, Mr. M.	1999 – 2002
McLaughlin, Dr. J.	1998 – 2001
Misener, A.	2003 – 2008
Morin, B.	2014 – Present
Murdoch, K.	2014 – 2017
Neill, Dr. R.	1991 – 2003
Paulin, M.	2007 – 2014
Picot, Dr. J.E.	1970 – 1976
Poirier, Dr. S.R.	2003 – 2008
Pottie, Dr. R.	1982 – 1990
Reeder, K.	2007 – 2014, Chairperson 2008 – 2014
Rinehart, Dr. S.	2016 – Present, Chairperson 2016 – Present
Ritchie, A.	2003 – 2008
Robinson, Dr. D.B.W.	1962 – 1971
Rogers, D.	2016 – Present
Roy, Mr. R.	1976 – 1979
Ryan, W.	2003 – 2004
Savoie, Dr. M.A.	1962 – 1976
Schnarr, Mr. J.	1988 – 1991
Seagrave, M.	2016 – Present
Shemilt, Dr. L.W.	Chairman 1962 – 1969, Member 1962 – 1980
Simmonds, K.	2016 – Present
Stafford, Mr. H.	1972 – 1975
Swift, Mr. F.	1988 – 1991
Travis, Dr. M.E.	1985 – 1988

Member (cont'd)	**Years Served** (cont'd)
Valenta, Mrs. N.	1976 – 1979
Verret, M.	1988 – 1991
Volpé, G.	2003 – 2008
Vo-Van, Dr. T.	2003 – 2003
Walton, Mr. F.T.	1967 – 1971
Watson, Ms. K.	1991 – 1997
Webster, Mr. D.	1975 – 1987
White, Mr. W.H.	1967 – 1971
Wilson, Dr. F.	1991 – 1998
Young, Mr. A.	1981 – 1987

Total number: 119

Mean years in position: 5.7

Longest serving Board Member: Dr. Ken V. Cox (29 years),

Longest serving Board Chairperson: Dr. Knut Grotterod (22 years as Chair out of 23 years as Member)

Figure 8.1. RPC's Board of Directors in 2019. Top row, L-R: Annette Comeau, Janet Gagnon (Vice-Chair), Dr. Levi Hargrove, Jeff Jennings. Middle row, L-R: Cathy Larochelle, Thomas MacFarlane, Irene McCardle, Bernard Morin. Bottom row, L-R: Dr. Shelley Rinehart (Chairperson), David Rogers, Meaghan Seagrave, Kelli Simmonds.

Appendix 8.3. RPC Executives and Senior Management Team

RPC has had five official Executive Directors. When RPC was first created, Mr. D.W. Gallagher served unofficially as Executive Director for a few months, prior to the appointment of Mr. F.R. Drummie as the first official, Order-in-Council appointed Executive Director [32]. During Eric Cook's tenure, the title was changed to Executive Director/CEO.

Executive Directors	**Years Served**
Mr. F.R. Drummie	1962 – 1963
Dr. Claude Bursill	1963 – 1983
Dr. Roy S. Boorman	1983 – 1992
Dr. Peter Lewell	1992 – 2004
Mr. Eric Cook	2004 – Present
Total number:	5
Mean years in position:	11.4 years
Longest serving:	Dr. Bursill, with 20 years

Working with and for the Executive Director/CEO are an Executive Management Team (see Figure 8.2) and a Senior Management Team (see Figure 8.3).

Figure 8.2. RPC's Executive Management Team in 2019. Clockwise from top left: Dr. Diane Botelho, Chief Science Officer; Bev Corey, Controller; Shelley Janes, Director of Human Resources; Steve Holmes, Chief Operating Officer; and, (at centre) Eric Cook, Executive Director/CEO.

Figure 8.3. RPC's Senior Management Team in 2019. Clockwise from top left: John Aikens, Director of Engineering Services; Matt Ashfield, Director of IT; Leo Cheung, Director of Process Engineering; Dr. Ben Forward, Director of Applied and Experimental Bioscience; Ross Kean, Director of Inorganics; and, Bruce Phillips, Director of Organics.

Appendix 8.4. RPC Subsidiaries.

Name	History
Innovent Ltd.	o Federally incorporated November 1973 o Purpose: to stimulate PTPP activities (other than pottery – *see* Galleon Ware Ltd.) o Went inactive ca. October 1976 o Dissolved March 1997
Galleon Ware Ltd.	o Incorporated in N.B. April 1975 o Purpose: to stimulate PTPP activities in pottery and related products o Held the Galleon Ware trademark and marketed and sold pottery commercially o Went inactive ca. July 1977 o Wound-up and assets sold in October, 1977
Minuvar Ltd.	o Incorporated in N.B. April 1975 o Purpose: mineral resource development, including limestone and manganese deposits, and also soil remediation. o Still held but dormant at present
Enhanced Recovery Systems Ltd. (ERS)	o Incorporated in N.B. September 1983 o Purpose: pilot-plant-scale development, testing, and demonstration of mineral processing technologies including RPC's Sulfation-Roast-Leach (SRL) process o Currently inactive

Figure 8.4. Example of the product lines from RPC's Galleon Ware subsidiary, *circa* 1977.

Appendix 8.5. RPC Joint-Ventures[64].

Name and Inception	Nature
Atlantic Provinces Institute of Forestry, 1964	o Joint with UNB and other agencies, o Purpose: To encourage higher productivity of utilization of forest resources. See Chapter 2.
Manufacturing Technology Centre (MTC), 1985	o Joint with UNB, Université de Moncton, and 3 community colleges, o Purpose: To transfer CAD/CAM technology to N.B. companies. See Chapter 4.
Incutech Brunswick Inc., 1988 – 1992	o Joint with UNB, o Purpose: To support entrepreneur-led technology-intensive manufacturing businesses in N.B. See Chapter 4.
Centre for Nuclear Energy Research (CNER), 1992 - 2010	o Joint with UNB and NB Power, o Purpose: To conduct R&D related to the safety and operation of nuclear generating plants. See Chapter 4.

---

[64] Joint ventures meaning jointly founded and operated in practice, and not necessarily meeting the definition of legal joint-ventures.

Appendix 8.6. RPC Acquisitions.

Year	Program or Organization	From	References
1963	Technical Information Service (TIS) program in New Brunswick	National Research Council	[38]
1964	Technical Information Service (TIS) program in Prince Edward Island	National Research Council	[38]
1974, 1975	Numerous limestone mining claims were staked by RPC subsidiary Minuvar Ltd. in the Elmtree and LaPlante areas	N.B. Department of Natural Resources and Energy	[127]
1976	Numerous abandoned manganese mining claims were acquired by RPC subsidiary Minuvar Ltd. in the Woodstock area	N.B. Department of Natural Resources and Energy	[128-130]
2011	Environmental analytical laboratories in Moncton.	Caduceon Environmental Laboratories	[254]
2016	Provincial dairy, fish health, and environment labs	Province of New Brunswick, Fredericton	[271,272]
2016-2017	Toxicology lab	Buchanan Environmental Ltd., Fredericton	[273,274]

Appendix 8.7. RPC Merit Awards.

The ***RPC Merit Award***, instituted in 1999/2000, is presented annually to an individual or team of individuals who have made an outstanding contribution or had a substantial achievement for RPC. Each year's award team is highlighted in RPC's annual report. For consistency, the following excerpts have been slightly edited from the originally published stories.

The *1999-2000* award was presented to the Mechanical Systems & Diagnostics Department in recognition of their excellent achievement in developing a prototype "X-brace" folding wheelchair enabling RPC's client, a small, innovative firm in New Brunswick to patent and commercialize its product idea. ***Team members:*** John Aikens, Brian Bell, Greg Brown, John King, Melanie Lalonde, Marc Robichaud, and Troy Young.

The *2000-01* award was presented to the ISO Implementation Team, comprised of the ISO Committee and ISO Intemal Auditors, led by Geri Tees, Quality Assurance Coordinator, for their excellent achievement in obtaining ISO 9001:2000 registration for RPC by diligently working with, encouraging, and inspiring their colleagues toward this goal. **Team members:** Dr. Bev Bacon, John Capar, Peter Crowhurst, Tim Doherty, Steve Fox, Ross Kean, Rick Kowalski, Melanie Lalonde, Melanie Mathews, Mike Miller, Chris Riley, Chris Steeves, and Geri Tees.

A *2001-02* award was presented to the Mineral Processing Group in recognition of their outstanding achievement in the development of a bacterial leaching process for extracting copper and nickel from low-grade ores. The process has been successfully piloted in two 5,000 tonne scale heaps as the RPC/Titan Bioheap™ Bacterial Leaching Process by the client Titan Resources NL of Perth, Australia. **Team members:** Lisa Baker, Leo Cheung, Ross Gilders, and Behnam Yazdani.

A separate *2001-02* award was presented to John Capar and Chris Steeves of the Physical Metallurgy Department for their outstanding achievement in, respectively, advancing guided wave inspection to enable crack detection in feeder bends of CANDU reactors, and developing procedures for inspecting suction rolls in the pulp and paper industry under critical conditions. Both have become recognized as experts in their respective fields.

The *2002/03* award was presented to Dr. Rachael Ritchie and other members of RPC's Molecular Biology Group in recognition of their outstanding work in developing RPC's molecular genetics expertise. RPC has become an important technical resource in genetics management for New

Brunswick's aquaculture industry, assisting the industry to access new opportunities and to discover solutions to old problems. ***Team members:*** Marcia Cook, Dr. Steve Griffiths, Eric Johnsen, Rebecca Liston, Erin MacDonald, Dr. Dougie McIntosh, Diane Moores, Dr. Rachael Ritchie, Nathalie Simard, and Sherry Vincent.

The ***2003/04*** award was presented to Dr. Pawel Kielczynski and members of RPC's Mechanical Systems and Diagnostics Department in recognition of their outstanding work in developing and bringing to market RPC's Windows©-based, high-temperature online temperature compensated ultrasonic flowmeter. ***Team members:*** John Aikens, Brian Bell, Dr. Pawel Kielczynski, and Marc Robichaud.

The ***2004/05*** award was presented to the Laboratory Information Management System (LIMS) software development team led by Sue Harris, Programmer Analyst, in recognition of their outstanding achievement with the development and implementation of RPC's proprietary LIMS. This system has had a substantial impact on the efficiency and the quality of RPC's client service and formed the foundation for further improvement on an ongoing basis. The LIMS achievement is an excellent example of teamwork with multiple departments and with multiple disciplines. It also emphasized RPC's commitment to improving effectiveness and quality and competing globally in delivering value to RPC's clients. ***Team members:*** Angela Colford, Peter Crowhurst, Lenora Fanjoy, Sue Harris, Ross Kean, Bruce Phillips, and Sandi Walker.

The ***2005/06*** award was presented to the RPC dioxin team for their exemplary work in analyzing samples from Canadian Forces Base Gagetown for dioxins and furans. This effort was in support of the fact-finding mission investigating concerns about Agent Orange and other potentially harmful chemicals previously used at the base. After winning a national competition for the project, the urgency for facts increased and the project time frame compressed dramatically from two years to five months. The team responded to the challenge developing an approach to more than double their capacity while complying with stringent quality control requirements. The project finished ahead of schedule and the client was delighted with RPC's efforts. This is an excellent example of RPC's unique ability to combine world-class science with unsurpassed customer service. At the project close-out meeting, the client's distinguished dioxin specialist recognized his confidence in RPC's data and noted they could not have received higher quality data from any commercial lab in Canada. ***Team Members:*** Lynn Jewett, Dr. John Macaulay, Sean McGrath, Stacey Munn, Josh Perry, Troy Smith, Mike Skinner, and Kent Walsh.

The *2006/07* award was presented to the members of the mineral processing team of RPC's Process and Environmental Technology department for their exemplary work in advancing RPC's mineral extraction capability in flotation and applying the technology to problematic polymetallic ore from the Aguas Teñidas ore-body in Spain. RPC's success allowed its client to proceed with the development of a mine to be fully operational in 2008 generating follow-on work and new inquiries for RPC's flotation expertise. This project is an excellent example of New Brunswick-based expertise being globally competitive. In this case, flotation and mineral extraction expertise developed to support NB-based industry was exported to Spain. Furthermore, the project incorporated Spanish regulations that prohibit the use of cyanide, a commonly employed chemical in the mining industry. RPC's client proceeded with the full development of the mine and contracted RPC for further development (see *"Reign in Spain"* in Chapter 6). **Team Members:** Leo Cheung, Katie Cougle, Ross Gilders, Jodi Jewett, Keith McLellan, Matt Ness, Craig Steeves, and Sandi Walker.

The *2007/08* award was presented to the members of the non-destructive testing (NDT) team for their outstanding achievement in developing NDT solutions for remote areas of a nuclear reactor. The first project involved development of a magnetic rubber replica kit for surface crack inspection in feeder pipes for both NB Power and Hydro Quebec. This project included the complete development of the inspection technique, design and construction of specialized rubber injection system, clamp-on molds with air bladders, and a special miniature magnetic yoke and power box. This technique and tooling is now the accepted CANDU Operator's Group method to detect and archive tiny surface cracks on the outer diameter of feeder bends. The second project was the development of remote magnetic particle and radiographic inspection tools for AECL. These tools were developed to complete single feeder replacements in very difficult to access areas of the reactor (see *"Powerful Innovations"* in Chapter 6). This project success is representative of the type of service RPC provides. A critical need was identified by a client, a motivated team was assembled, and innovative methods were developed to deliver a solution to the client. **Team Members:** John Aikens, Brian Bell, John King, Melanie Lalonde, Chris Steeves, Tony Wilkins, and Troy Young.

The *2008/09* award was presented to the members of the Food, Fisheries and Aquaculture (FFA) Department, fish health team for their outstanding achievement conducting applied research to open export markets for a New Brunswick client. The Chilean salmon aquaculture industry was plagued with infectious salmon anaemia and pancreas disease, and the client saw a business opportunity in providing pathogen-free eggs to the Chilean industry. With a very tight schedule, the client engaged FFA's fish health group to conduct a custom designed fish health screening of brood stock fish to allow export of fish eggs to Chile. This was the first time anyone in Canada had attempted to ship eggs to Chile meaning RPC staff had to work closely with the client, Canadian Food Inspection Agency (CFIA) personnel and Chilean regulators. The work necessitated rapid refinement of new real-time assays for the infectious salmon anaemia virus (ISAV) and Salmon alphavirus (SAV) pathogens and a microscope-based assay for *Piscirickettsia salmonis*. The research also resulted in development of a state-of-the-art, highly sensitive assay for ISAV which provided the client more detailed analysis of their fish. The real-time RT-PCR assay developed within this project has become a regular service offering in FFA and resulted in significant new work for RPC while helping protect New Brunswick's aquaculture industry. **Team Members:** Janice Comeau, Lenora Fanjoy, Dr. Ben Forward, Eric Johnsen, Rebecca Liston, Dr. Tony Manning, Dr. Rachael Ritchie.

The *2009/10* award was presented to Geri Tees and Sue Harris in recognition of their successful development of a Navision-based document control system. The system allows for the electronic review and approval of standard operating procedures, greatly reducing effort required for managing the document approval process. RPC has realized many benefits from this success, including: a significant reduction in time spent printing, circulating and tracking controlled copies of documentation, a significant reduction in time obtaining document approvals and evidence of document review and reading, increased assurance that the document in use is the current version -the copy available through Navision is always the latest approved copy available, and easy accessibility to documented procedures throughout the building, either for reference or for training purposes. This system has, and will continue to, vastly improve the ease and quality of document maintenance, and have a favourable impact on RPC's performance during accreditation and certification audits. **Team Members:** Sue Harris and Geri Tees.

The **2010/11** award was presented to the team who developed and launched RPC's accredited genetics testing service line including the www.rpcgentics.com website. This project evolved from RPC's DNA capabilities in the aquaculture and wildlife forensics sector. Research was completed to understand market demand and the requirements to obtain accreditation for human genetics testing, specifically paternity and maternity analysis. Technical requirements, including state-of-the-art DNA extraction capability, were established. Standard operating procedures were developed then audited and approved by the Standards Council of Canada. A consumer-friendly website was developed that allows for online ordering of test kits and secure payment with credit card or PayPal. The service was officially launched in January 2011 and has resulted in a variety of projects including the identification of siblings and parents, resolution of inheritance disputes, and assisting provincial medical examiners to identify unknown remains. The development and launch of new services is crucial to RPC's long-term success. This highly technical scientific service was combined with an easy-to-use customer interface and immediately resulted in international business. The project is an important achievement and an excellent example of RPC's renewal. **Team Members:** Samantha Atkinson, Sherry Binette, Amy Brown, Lenora Fanjoy, Dr. Ben Forward, Eric Johnsen

The **2011/12** award was presented to the Mechanical Systems and Diagnostics team, who developed a system for measuring feedwater flow and temperature in a nuclear power plant. This effort required the design, build, test, and calibration of a complete turnkey system including ultrasonic transducers, brackets, software, and cabling. RPC utilizes the innovation for making high temperature flow and temperature measurements at nuclear power facilities in Ontario. These measurements are critical to safe and efficient reactor operation. This is an excellent example of market-led research and development. A client had a technical challenge and RPC conducted research to develop an innovative solution. **Team Members:** John Aikens, Brian Bell, Pat Hudson, Dr. Pawel Kielczynski, Robert Kirouac, John King, and Troy Young.

The **2012/13** award was presented to the team from the Food, Fisheries and Aquaculture (FFA) department for the project: "Analysis of Male-Specific Coliphage (MSC), a new tool for the assessment of wastewater treatment efficacy." MSC is a virus commonly found in sewage effluent and can be used as a viral indicator of human fecal pollution in water and shellfish.

RPC implemented several methodologies for growth and analysis of MSC following inquiries from federal and provincial departments. These methodologies formed the basis for eight projects that assessed wastewater treatment efficacy from sewage treatment plants, fish processing facilities, as well as a federal quarantine research station. Results have contributed to the reassessment of restricted shellfish harvesting boundaries, permitted industry compliance with new regulations, and enabled cutting-edge product improvement research. This project represents a new revenue stream and is an excellent example of one that applies science to solve industry needs. **Team Members:** Lenora Fanjoy, Dr. Ben Forward, Renée Jeffrey, Eric Johnsen, Jessica Jones, Rebecca Liston, Dr. Tony Manning, and Dr. David Thumbi.

The **2013/14** award was presented to the Air Quality group for their efforts in significantly improving the breathing air lab with increased capacity, improved procedures, expanded scope and a significantly increased client base. Additionally, electronic reporting for asbestos and spore traps was implemented. Their efforts included upgrading of equipment with improved capacity, and expansion of test parameters. The dew-point measurement process was completely revamped, leading to improved quality. Two important new accreditations were added; methane in water, and expanded parameters for divers' breathing air. The group's efforts to implement electronic reporting continued with the addition of spore traps and asbestos. All of the above led to new clients, expanded capacity, and expanded capability - in one case providing critical assistance to a client who had a sudden need for large throughput. Improved quality, improved customer service and expanded business made this effort worthy of the award. **Team Members:** Dr. Diane Botelho, Bryan Bourque, Jodi Buckingham, Erin Craig, Karla McLellan, Darren Tarr, and Sandi Walker.

The **2014/15** award was presented to the team led by the Organic Analytical Services department for their efforts in creating a service line in response to client inquiries resulting from the newly created Marijuana for Medical Purposes Regulations. Through these new regulations Health Canada now licenses private companies to grow and sell cannabis to people possessing valid prescriptions. Mandatory quality testing is required as part of the new regulations. This was an excellent opportunity for several departments within RPC to work together to supply a service to a new client base. There were three departments that were challenged to validate new methods or existing methods to prove they were fit for purpose when testing medicinal cannabis. RPC also upgraded its existing license to accommodate the increased amount of controlled substance that would be handled. RPC's ability to offer these services provided its clients an effective laboratory partner that produces quality results in a timely manner at competitive prices. The project success is a good example of RPC's agility in recognizing an industry need and proactively acting to serve their need. The project is an excellent example of the benefits of multidisciplinary teamwork. It resulted in new clients, expanded capacity, and expanded capability while helping the industry to grow and, therefore, was worthy of the award. **Team Members:** Karen Broad, Peter Crowhurst, Cathy Hay, Bruce Phillips, and Troy Smith.

The **2015/16** award was presented to the team who developed and manufactured an Automated Laboratory Workcell[65]. This project is an excellent example of how RPC can complete complex objectives utilizing modern technology and a cross-functional team including close collaboration with the client. The challenge was to automate a labour-intensive biotech process within a tightly controlled environment. This was not a standard automation project. The solution required complex research, advanced industrial design, robotics, and machine vision. The Engineering Services

---

[65] Project details have been excluded to protect the resulting intellectual property.

team developed the complete process, detailed the design, manufacturing and testing of the prototype machine. After successful testing, operator training, machine installation and commissioning at the client's site were completed. The prototype was assembled and put into full production in December of 2015, exceeding targeted first-year production rates. The RPC team delivered an ideal example of industry-driven innovation. **Team Members:** John Aikens, Brian Bell, Chris Davenport, Dr. Ben Forward, Steve Holmes, John King, Melanie Lalonde, Scott Sanford, and Troy Young.

The **2016/17** award was presented to the team working on the development of rubber devulcanization technology with Rubreco. The project began with an independent evaluation of Rubreco's patent to devulcanize rubbers from the tire industry and progressed to bench-top testing, proof of concept, through to development of a full-scale production unit and optimization. Rubreco has a series of eight patents resulting from the work. The Rubreco technology takes crumb rubber from car tires, truck tires, and tire liners, and devulcanizes it in the Rubreco reactor using only water as a solvent. The devulcanized rubber can then be used to supplement new rubber. This is clean technology that is both environmentally-friendly and economical. This breakthrough innovation is an environmentally friendly solution for the significant used-tire challenge. **Team Members:** Brian Bell, Neri Botha, Jodi Buckingham, Leo Cheung, Katie Cougle, Alex Kyle, Keith McLellan, Andrew Mullin, Matthew Ness, Darren Tarr, and Sandi Walker.

The **2017/18** award was presented to the employees associated with the launch of RPC Quality Team and the successful achievement of ISO 9001:2015. This project is an excellent example of teamwork as it included members from each of RPC's operating departments. The team was assembled, received training and was tasked with leading the effort to help RPC become ISO 9001:2015 certified. The 2015 accreditation required significant updates to the ISO 9001 standard including new elements. Documentation was prepared, procedures deployed and processes internally audited leading up to the successful external audit in March of 2018. RPC is now ISO 9001:2015 registered. This team approach is consistent with RPC's quality philosophy in that quality is everyone's business. **Team Members:** Chadwick Anderson, Lisa Banks, Neri Botha, Bryan Bourque, Sara Cockburn, Jennifer Doucette, Lisa Ferrish, Angie Guitard, Melanie Lalonde, and Gillian Travis.

The **2018/19** award was presented to the employees associated with building RPC's cannabis business. Building on decades of analytical science experience for the agriculture, aquaculture, forestry, food, and environmental sectors, RPC has become a leading provider of cannabis science services. The leadership has resulted in significant accomplishments including growth in revenue, growth in employment, adding new clients, and helping RPC and

New Brunswick to be recognized nationally and internationally. During the past year, new labs, new equipment, and new quality procedures were advanced. RPC was recognized by Lift & Co. as the Top Testing lab in Canada. The efforts have stimulated growth in other departments and new applied research services. Most importantly, this has helped dozens of clients to grow and create thousands of jobs. **Team Members:** April Boudreau, Karen Broad, Bruce Phillips, and Troy Smith.

> ***RPC Trivia:*** Numerous employees have received the RPC Merit Award multiple times over the years, including ten employees that have won it at least four times. Of these, John Aikens and Melanie Lalonde have each won it five times, and ***Brian Bell holds the record, having won the award an amazing six times!***

Appendix 8.8. Patents Issued on RPC Staff Inventions.

As is the case with technical presentations, publications, and books (Appendix 8.6), with such a strong focus on helping business, industry, and communities grow and thrive, and because so much of RPC's work is done under proprietary contracts, opportunities for RPC employees to get their inventions patented are much reduced compared with the academic, government, or even the industrial sectors. In many cases, intellectual property (IP) rights from RPC's work vest with the clients, and in quite a few cases any new IP is protected as trade secrets.

The following list of issued patents with RPC inventors is almost certainly incomplete, particularly since for some inventions patent applications are sought in a large number of countries worldwide. The list does, however, provide some examples of areas in which RPC has demonstrated an unusually high degree of inventive skill.

The complete patents are normally available for free from the issuing countries. For Canadian and U.S. patents, for example, see:

- o http://www.ic.gc.ca/opic-cipo/cpd/eng/
- o http://patft.uspto.gov/

1960s:
- o Wilstead, G.H., Gallop, R.A., Chan, M.S., "Texture Measuring Instrument," Canadian Patent 787,637, June 18, 1968.

1970s:
- o Brown, G.D., Sprague, J.B., "Use of Sequestering Agents," Canadian Patent 868,648, April 13, 1971.
- o Brown, G.D., "Process for Preparing Quick Cooking Rice," Canadian Patent 873,808, June 22, 1971.
- o Wilkomirsky, I.A.E., Boorman, R.S., Salter, R.S., "Recovery of Non-Ferrous Metals by Thermal Treatment of Solutions Containing Non-Ferrous and Iron Sulphates," World Patents:
  - o Australian Patents AU4608479 (A), November 8, 1979; AU526363 (B2), January 6, 1983,
  - o Belgian Patent BE876054 (A), September 3, 1979,
  - o U.K. Patents GB2020261 (A), November 14, 1979, GB2020261 (B), November 14, 1979,
  - o German Patent DE2914823 (A1), November 15, 1979,
  - o Finland Patents FI791433 (A) November 6, 1979; FI69104 (B), August 30, 1985 ; FI69104 (C), December 10, 1985,
  - o French Patent FR2424962 (A1), November 30, 1979;

FR2424962 (B1), July 22, 1983,
- o Irish Patents IE790888 (L), November 5, 1979; IE48112 (B1), October 3, 1984,
- o Japan Patents JPS54147102 (A) November 17, 1979; JPS63494 (B2), January 1, 1988.
- o South African Patent ZA7901583 (B), May 28, 1980.

1980s:

- o Wiesner, C.J., "Process for Preparing 11-Tetradecenal Insect Pheromone," Canadian Patent 1,077,059, May 6, 1980.
- o Wiesner, C.J., "Process for Preparing Insect Pheromones," U.S. Patent 4,212,830, July 15, 1980.
- o Wilkomirsky, I.A.E., Boorman, R.S., Salter, R.S., "Recovery of Non-Ferrous Metals by Thermal Treatment of Solutions Containing Non-Ferrous and Iron Sulphates," U.S. Patent 4,224,122, September 23, 1980.
- o Wilkomirsky, I.A.E., Boorman, R.S., Salter, R.S., "Recovery of Non-Ferrous Metals by Thermal Treatment of Solutions Containing Non-Ferrous and Iron Sulphates," U.S. Patent 4,317,803, March 2, 1982.
- o Wilkomirsky, I.A.E., Boorman, R.S., Salter, R.S., "Recovery of Non-Ferrous Metals by Thermal Treatment of Solutions Containing Non-Ferrous and Iron Sulphates," Canadian Patent 1,121,605, April 13, 1982.
- o Salter, R.S., Boorman, R.S., Gilders, R.D., "Recovery of Lead and Silver from Minerals and Process Residues," U.S. Patent 4,372,782, February 8, 1983.
- o Salter, R.S., Boorman, R.S., Gilders, R.D., "Process for the Recovery of Lead and Silver from Minerals and Process Residues," Canadian Patent 1,156,048, November 1, 1983.
- o Wilkomirsky, I.A.E., Boorman, R.S., Salter, R.S., "Recovery of Non-Ferrous Metals by Thermal Treatment of Solutions Containing Non-Ferrous and Iron Sulphates," U.S. Patent 4,415,540, November 15, 1983.
- o Salter, R.S., Boorman, R.S., Wilkomirsky, I.A.E., "Process for the Recovery of Non-Ferrous Metals from Sulphide Ores and Concentrates," Canadian Patent 1,203,083, April 15, 1986.
- o Salter, R.S., Boorman, R.S., Wilkomirsky, I.A.E., "Process for the Recovery of Non-Ferrous Metals from Sulphide Ores and Concentrates," U.S. Patent 4,619,814, October 28, 1986.
- o Wiesner, C.J., "Skunk Odor Shampoo," U.S. Patent 4,834,901, May 30, 1989.
- o Theriault, M., "Dewiring Tool," U.S. Patent 4,841,619, June 27, 1989.

- o Desjardins, C.D., Sharifian, H., MacLean, G.K., "Lithium-Lithium Nitride Anode," U.S. Patent 4,888,258, December 19, 1989.

1990s:

- o Wiesner, C.J., "Skunk Odour Shampoo," Canadian Patent 1,286,230, July 16, 1991.
- o Desjardins, C.D., Sharifian, H., MacLean, G.K., "Lithium-Lithium Nitride Anode," Canadian Patent 1,302,491, June 2, 1992.
- o Theriault, M., "Dewiring Tool," Canadian Patent 1,315,757, March 2, 1993.
- o Godin, G., Lambert, P.A., "Wheel assembly for a wheelchair, incorporating a change speed hub," U.S. Patent 5,486,016, Jan. 23, 1996.
- o Meier, H.J., Landry, G., Caissie, R., "Method and Material for Extending the Shelf-Life of Fresh Foods," Canadian Patent 2,135,416A1, Feb. 10, 1998.

2000s:

- o Simard, N.C., Brouwers, H., Jones, S., Griffiths, S., Valenzuela, P., Burzio, L., "Sequences from *Piscirickettsia salmonis*," U.S. Patent 6,887,989, May 3, 2002.
- o Meier, H.J., Landry, G., Caissie, R., "Packaging Material for Curing or Marinating Fresh Foods During Storage at Low Temperatures," U.S. Patent 6,623,773, Sept. 23, 2003.
- o Griffiths, S., Ritchie, R.J., Heppell, J., "Nucleic Acid and Amino Acid Sequences of Infectious Salmon Anaemia Virus and Their Use as Vaccines," 3 U.S. Patents: 6,919,083, July 19, 2005; 7,128,917, October 31, 2006; 7,199,108, April 3, 2007.
- o Hunter, C.J., Williams, T.L., Purkiss, S.A.R., Cheung, L.W., Connors, E., Gilders, R.D., "Bacterial Oxidation of Sulphide Ores and Concentrates," U.S. Patent 7,189,527, March 13, 2007.
- o Griffiths, S., Ritchie, R.J., "Nucleic Acid and Amino Acid Sequences of Infectious Salmon Anaemia Virus and Their Use as Vaccines," U.S. Patent 7,201,910, April 10, 2007.
- o Griffiths, S., Ritchie, R.J., Simard, N.C., "Hsp60 from Arthrobacter," U.S. Patent 7,297,783, November 20, 2007.
- o Aikens, J., "Device and System for Corrosion Detection," U.S. Patent 7,552,643, June 30, 2009.

2010s:

- o Griffiths, S., Ritchie, R.J., Simard, N.C., "Hsp70 from Arthrobacter," U.S. Patent 7,674,892, March 9, 2010.

o Griffiths, S., Ritchie, R.J., Simard, N.C., "Hsp60 from Arthrobacter," U.S. Patent 7,803,924, September 28, 2010.

o Simard, N.C., Brouwers, H., Jones, S., Griffiths, S., Valenzuela, P., Burzio, L., "Sequences from *Piscirickettsia salmonis*," U.S. Patent 7,842,296, November 30, 2010.

o Griffiths, S., Ritchie, R.J., Heppell, J., "Nucleic Acid and Amino Acid Sequences of Infectious Salmon Anaemia Virus and Their Use as Vaccines," U.S. Patent 7,998,484, August 16, 2011.

o Harrison, B., Hooper, H., Gilders, R., Cheung, L., Ness, M., "Separating Devulcanized Rubber," U.S. Patent 8,415,402, April 9, 2013.

o Simard, N.C., Brouwers, H., Jones, S., Griffiths, S., Valenzuela, P., Burzio, L., "Fish Vaccine Against *Piscirickettsia salmonis*," Canadian Patent 2,402,718, June 25, 2013.

o Cheung, L., Harrison, B., Hooper, H., Gilders, R., Ness, M., "Separating Devulcanized Rubber," Canadian Patent 2,742,135, October 1, 2013.

o Griffiths, S., Ritchie, R.J., "Nucleic Acid and Amino Acid Sequences of Infectious Salmon Anaemia Virus and Their Use as Vaccines," Canadian Patent 2,378,437, January 21, 2014.

o Hunter, C.J., Williams, T.L., Cheung, L.W., Connors, E., Gilders, R.D., Purkiss, S.A.R., "Improved Bacterial Oxidation of Sulphide Ores and Concentrates," European Patent EP 2,224,022 B1, filed August 29, 2000.

o Rogers, D.M., Aikens, J., "Root Vegetable Product," 28 U.S. Design Patents: D707,418, June 24, 2014; D712,618, September 9, 2014; D714,017, September 30, 2014; D714,517, October 7, 2014; D714,521, October 7, 2014; D720,916, January 13, 2015; D720,917, January 13, 2015; D721,217, January 20, 2015; D721,218, January 20, 2015; D721,219, January 20, 2015; D721,220, January 20, 2015; D722,417, February 17, 2015; D722,418, February 17, 2015; D722,419, February 17, 2015; D725,343, March 31, 2015; D728,892, May 12, 2015; D728,894, May 12, 2015; D729,491, May 19, 2015; D729,492, May 19, 2015; D735,441, August 4, 2015; D742,091, November 3, 2015; D744,716, December 8, 2015; D745,244, December 15, 2015; D745,245, December 15, 2015; D746,019, December 29, 2015; D748,366, February 2, 2016; D748,891, February 9, 2016; D762,341, August 2, 2016.

o Griffiths, S., Simard, N.C., Ritchie, R.J., "Hsp70 from Arthrobacter," Canadian Patent 2,492,556, September 16, 2014.

o Griffiths, S., Ritchie, R.J., Simard, N.C., "Hsp60 from Arthrobacter," Canadian Patent 2,513,061, December 12, 2015.

o Harrison, B., Hooper, H.A., Ness, M., "Dewatering Devulcanized Rubber," U.S. Patent 10,087,301, October 2, 2018.

Pending (this list is incomplete, but illustrative of RPC patents pending):

o Gilders, R., Desjardins, C.D., "Whole Ore Treatment Process for the Recovery of Nickel, Cobalt and Copper from Sulfide Ore and Tailings," Canadian Patent Application 2,064,543, filed March 31, 1992.

o Hunter, C.J., Williams, T.L., Cheung, L.W., Connors, E., Gilders, R.D., "Improved Bacterial Oxidation of Sulphide Ores and Concentrates," World Patent Application WO 01/18264 A1, filed August 29, 2000.

o Aikens, J.W., Rogers, D.M., "Blade Assembly and Food Cutting Device Incorporating the Same," U.S. Patent Applications 14/242,232 and 15/454,552, filed April 1, 2014.

o Rogers, D.M., Aikens, J.W., Rincon, C., "Rotary Blade Assembly for Cutting a Food Product Into Helical Strips," U.S. Patent Application 14/459,854, filed August 14, 2014.

o Rogers, D.M., Aikens, J.W., Rincon, C., "Device and Method for Cutting Food Pieces Into a Twisted Shape," U.S. Patent Application 14/515,107, filed October 15, 2014.

o Rogers, D.M., Aikens, J.W., Rincon, C., "Wiggle-Shaped Food Strip and Method of Cutting the Same," U.S. Patent Application 14/576,375, filed December 19, 2014.

o Rogers, D.M., Aikens, J.W., Rincon, C., "System and Method for Cutting Hasselback Food Items," U.S. Patent Application 14/597,840, filed January 15, 2015.

o Adams, G.W., McCartney, A.W., Lawless, J.F., Aikens, J., Davenport, C., "Systems, Methods and Apparatuses for Processing Plant Embryos," U.S. Patent Application 15/847,397, filed December 19, 2017.

o Adams, G.W., McCartney, A.W., Aikens, J., Davenport, C., MacLean, M., "Systems, Methods and Apparatuses for Processing Seedlings," U.S. Patent Application 16/227,903, filed December 20, 2018.

Appendix 8.9. Canada's Provincial Research Councils and Selected Other Canadian Research and Technology Organizations (RTOs).

Established	Evolution
1906: Horticultural Experiment Station "Jordan Harbour" (Vineland Research Station)	Later renamed Horticultural Research Institute of Ontario, acquired by the University of Guelph in 1977, then spun-off as the not-for-profit Vineland Research and Innovation Centre in 2009.
1916: Honorary Advisory Council for Scientific and Industrial Research	Act of Parliament passed in 1917. From 1925, it became known under its "short title" as the National Research Council of Canada (NRC).
1921: Advisory Council of Scientific and Industrial Research of Alberta	Later renamed Science and Industrial Research Council of Alberta (SIRCA), Research Council of Alberta (RCA), Alberta Research Council (ARC), then became part of Alberta Innovates Technology Futures, then InnoTech.
1928: Ontario Research Foundation	A separate Ontario Research Council existed from the mid-1940s through 1955. ORF was renamed ORTECH Corp. in 1990, privatized in 1999; part became Process Research ORTECH Inc.
1930, 1947: Research Council of Saskatchewan	Re-established as the Saskatchewan Research Council (SRC) in 1947.
1944: British Columbia Research Council	Privatized as BC Research Corp. in 1988.
1946: Nova Scotia Research Foundation	Amalgamated into InNOVAcorp., a venture capital firm, in 1995.
1961, 2009: Newfoundland Research Council	1961 NF Research Council Act was passed but not implemented. The Research and Development Council was formed in 2009, renamed Research and Development Corp., and wound-up in 2018.
1962: New Brunswick Research and Productivity Council	Later renamed the Research and Productivity Council (RPC).
1963: Manitoba Research Council	Evolved into the Industrial Technology Centre (ITC, 1979) and what is now the Food Development Centre (1978).
1969 : Centre de Recherches Industrielle du Québec	Commonly known as CRIQ.
1985: L'Institut national d'optique	Commonly known as INO (National Optics Institute).
1987: PEI Food Technology Centre	Evolved into Bio|Food|Tech in 2011.
2007: FPInnovations	Created by amalgamating Forintek Canada Corp. (est. 1979), Forest Engineering Research Institute of Canada (FERIC, est. 1975), and the Pulp and Paper Research Institute of Canada (Paprican, est. 1925). FPI maintains major facilities in Quebéc and British Columbia.

Appendix 8.10. Associations of Provincial Research Councils.

NIRAC. The Non-Profit Industrial Research Association of Canada (NIRAC) was formed in 1969 to promote recognition of the existence, the value and the potential of non-profit industrial research organizations in Canada and to ensure for them a place in any national science policy. Members were the Nova Scotia and Ontario Research Foundations, the Research Councils of Alberta and Saskatchewan, New Brunswick Research and Productivity Council, and B.C. Research. In 1969 the provincial research councils had available to them about 670 000 square feet of laboratory, office, pilot plant and other space, and about $8 million worth of major equipment items [294]. Collaborative activities among the industrial research organizations under NIRAC seem to have rapidly diminished after 1972.

On the international scene, RPC was also one of the Founding Members of the World Association of Industrial and Technological Research Organizations (WAITRO) in October 1970 [295].

APRO. The provincial research organizations (PROs) began actively meeting together again in about 1983, and started using the name Association of Provincial Research Organizations (APRO). This relationship was formalized by the incorporation of The Association of Provincial Research Organizations Inc. in November 1984, although the short name APRO continued to be used for most purposes. The original APRO membership comprised the research councils of BC, Alberta, Saskatchewan, Manitoba, Ontario, Québec, New Brunswick, and Nova Scotia.

APRO meetings were commonly held in Ottawa to facilitate discussions with federal representatives, and APRO maintained a national office in Ottawa on April 1, 1990 [296]. The goal of APRO was to "*[use] science and technology for the economic and social development of Canada and the Provinces*" [297]. The APRO meetings gave the PROs "*an opportunity to exchange ideas among themselves and develop strategies, ... raise the level of awareness of the PROs and their capabilities to outsiders, ... [and] enable them to press their case jointly in connection with problems that may exist with policies or programs of federal departments ...*" [18].

In 1991 APRO became "APRO – The Canadian Technology Network" [298], and in about 1996 APRO was amalgamated with the Canadian Manufacturers' Association (CMA) and the Canadian Exporters' Association (CEA) to form the Alliance of Manufacturers & Exporters Canada, which in 2000 became Canadian Manufacturers & Exporters (CME).

I-CAN. After another lapse of nearly a decade, four[66] of the PROs began meeting together again in 2003. These four decided to re-create an

---

[66] Alberta Research Council, Saskatchewan Research Council, Industrial Technology Centre, and Centre de Recherche Industrielle du Québec.

association, which for a short time was referred to as the Alliance of Provincial Research Organizations, and under the older acronym APRO. The aims of the new alliance were to strengthen Canada's innovation performance as a means of maintaining and improving Canada's global competitiveness, and to help Canada's research and technology organizations (RTOs) come together to work on projects of regional and/or national interest that would enable industry to solve large technology problems and challenges that would be too broad, expensive or risky for any one individual RTO to address alone.

The Founding Directors were: John McDougall, Dr. Laurier Schramm, Dr. Denis Beaulieu, and Trevor Cornell, the first Members were ARC, SRC, ITC, and CRIQ, and the alliance was renamed Innoventures Canada Inc. (I-CAN™). I-CAN was incorporated in 2006. By 2015 more PROs had joined I-CAN[67]: FP Innovations (BC and Quebec), the National Optics Institute (INO, Quebec), Bio|Food|Tech (PEI), New Brunswick Research and Productivity Council (RPC), National Research Council (NRC), and Vineland Research and Innovation Centre (Ontario). This gave I-CAN coast-to-coast "reach" across Canada, with (collectively) over 6,000 employees, over \$1 billion/yr in market-oriented R&D services, and over 150 facilities across all ten provinces.

One of I-CAN's flagship programs has been the ***I-CAN™ Innovation School,*** which was developed to address the needs for education, training, and continuous professional development in technological innovation, research and technology organizations (RTOs), and in the components and the workings of effective innovation systems [299]. The first annual ***I-CAN™ Innovation School*** was launched in 2010, and it has continued in various forms to the present day.

By 2011 I-CAN members had collectively stimulated over \$1 billion in direct business investment in science and technology, created several billion dollars in economic benefits through technology commercialization, attracted significant foreign benefits, improved Canadian quality of life by mitigating or reducing environmental impacts, and provided scientific understanding for more effective government regulations [299].

---

[67] The Northern Centre for Advanced Technology (NORCAT) and Powertech Labs also joined but later left the alliance.

Appendix 8.11. Glossary and Acronyms.

ACOA Atlantic Canada Opportunities Agency.

Advisory Council of Scientific
And Industrial Research
of Alberta *See* InnoTech.

AECL Atomic Energy of Canada Ltd.

AI-TF *See* InnoTech.

Alberta Innovates –
Technology
Futures (AI-TF) *See* InnoTech.

Alberta Research
Council (ARC) *See* InnoTech.

ARC Alberta Research Council. *See* InnoTech.

BC Research
Corp. *See* British Columbia Research Council.

Bio|Food|Tech
Prince Edward Island (PEI) established the ninth provincial research council in Canada with the PEI Food Technology Centre, which was incorporated in 1987 as a wholly-owned subsidiary of Innovation PEI, a Provincial Crown Corporation. Its name was changed to Bio|Food|Tech in 2011.

British Columbia
Research
Council The fourth provincial research council to be established in Canada, it was created in 1944. It was privatized as BC Research Corp. in 1988.

CAD/CAM   (Computer-Aided Design, Computer-Aided Manufacturing) An RPC Manufacturing Technology Centre (MTC) was launched in 1985 to transfer CAD/CAM technology to N.B. companies. *See* Appendix 8.5 and Chapter 4.

CAEL   Canadian Association of Environmental Analytical Laboratories.

CANDU Nuclear
Reactor   **CAN**ada **D**euterium **U**ranium is the acronym for Canadian pressurized-heavy-water nuclear power reactors.

Centre de Recherches
Industrielle du
Québec   (CRIQ) The eighth provincial research council to be established in Canada, it was created in 1969.

Centre for Frontier Energy
Research   *See* C-FER Technologies.

C-FER Technologies
     Originally created in 1983 as Centre for Frontier Energy Research (C-FER), a not-for-profit research and engineering company. Its original mission was to focus on the development of Canada's Arctic and offshore oil and gas resources, but this was later changed to operational productivity in the conventional and unconventional oil industry. It was acquired by the Alberta Research Council in 1999, and made a subsidiary: C-FER Technologies (1999) Inc. It is currently a subsidiary of InnoTech.

CNC Machine   CNC refers to computer numerical control, meaning the automated control of (usually) machining tools.

CRIQ   *See* Centre de Recherches Industrielle du Québec.

CSA   Canadian Standards Association.

Enhanced Recovery
Systems Ltd.   (ERS) An RPC subsidiary created to focus on mineral processing opportunities. *See* Appendix 8.4.

ERS   *See* Enhanced Recovery Systems Ltd.

FERIC   *See* Forest Engineering Research Institute of Canada.

Food Development

Centre   *See* Manitoba Research Council.

Forest Engineering
Research Institute

of Canada   (FERIC) was a not-for-profit RTO focused on forest harvesting and transportation research, development, and testing. It was amalgamated into FPInnovations in 2007 (see below).

Forintek   Forintek Canada Corp. was a not-for-profit RTO focused on wood products research, development, and testing. It was amalgamated into FPInnovations in 2007 (see below).

FPI   See FPInnovations.

FPInnovations   (FPI) FPInnovations, a not-for-profit RTO, was formed in 2007 through an amalgamation of Forintek, FERIC, Paprican, and the Canadian Wood Fibre Centre.

Honorary Advisory Council for
Scientific and Industrial

Research   The original name for the National Research Council (NRC), which became better known under its newer name after 1925.

Horticultural Experiment

Station   *See* Vineland Research and Innovation Centre.

Horticultural Research
Institute of Ontario

                    (HRIO) *See* Vineland Research and Innovation Centre.

HRIO   Horticultural Research Institute of Ontario. *See* Vineland Research and Innovation Centre.

Industrial Research
Assistance

Program   (IRAP) *See* Technical Information Service.

Industrial Technology

Centre   *See* Manitoba Research Council.

InnoTech The first provincial research council to be established in Canada, it was created as the Advisory Council of Scientific and Industrial Research of Alberta in 1921, later renamed Science and Industrial Research Council of Alberta (SIRCA), then Research Council of Alberta (RCA), then Alberta Research Council (ARC), then Alberta Innovates – Technology Futures (AI-TF), the InnoTech (a subsidiary of Alberta Innovates).

INNOVAcorp *See* Nova Scotia Research Foundation Corp.

INO *See* L'Institut national d'optique.

ISO International Organization for Standardization.

IRAP *See* Technical Information Service.

ITC Industrial Technology Centre. *See* Manitoba Research Council.

L'Institut national d'optique
(INO, the National Optics Institute) A not-for-profit research and development company, INO was created in 1985 and focuses on optics and photonics research for industrial development purposes.

Manitoba Research
Council The seventh provincial research council to be established in Canada, it was created in 1963. It later evolved into the Industrial Technology Centre (ITC, 1979) and what is now the Food Development Centre (1978).

Manufacturing Technology
Centre (MTC) *See* CAD/CAM.

MEAP Mobile Energy Audit Program, also referred to as 'energy-bus' program.

MTC Manufacturing Technology Centre. *See* CAD/CAM.

National Optics
Institute *See* L'Institut national d'optique.

National Research Council

of Canada   (NRC) Canada's national research and technology organization, formed in 1916.

NBEPC   New Brunswick Electric Power Commission.

New Brunswick
Research and Productivity
Council   *See* Research and Productivity Council.

Newfoundland Research

Council   There was never a "Newfoundland Research Council" *per se*, as a Newfoundland Research Council Act was passed in 1961 but never implemented. *See, however,* Research and Development Corp.

Nova Scotia Research Foundation

 Corp.   (NSRFC) The fifth provincial research council to be established in Canada, it was created in 1946. It was amalgamated into INNOVAcorp., a venture capital firm, in 1995.

NRC   *See* National Research Council of Canada.

NSRFC   *See* Nova Scotia Research Foundation Corp.

Ontario Research

Foundation   (ORF) The second provincial research council to be established in Canada, it was created as the Ontario Research Foundation in 1927. A separate Research Council of Ontario existed from the mid-1940s through 1955 after which its activities were transferred to ORF. ORF was renamed ORTECH Corp. in 1990 and privatized in 1999.

ORF   *See* Ontario Research Foundation.

ORTECH
Corp.   *See* Ontario Research Foundation.

PAHs   Polyaromatic hydrocarbons.

PAMI   *See* Prairie Agricultural Machinery Institute.

Paprican   The Pulp and Paper Research Institute of Canada (Paprican) was a not-for-profit RTO focused on focused on pulp and paper research, development, and testing programs supported by industry and government members. It was amalgamated into FPInnovations in 2007 (see above).

PCBs   Polychlorinated biphenyl compounds.

PCP   Pentachlorophenol.

**PEI Food Technology
Centre**   *See* Bio|Food|Tech.

**Petroleum Recovery
Institute**   (PRI) The Petroleum Recovery Research Institute (PRRI) was incorporated as a not-for-profit research and engineering organization in 1966. Its mission was to focus on conventional oil recovery, particularly enhanced oil recovery (EOR) in Canada. In 1974, the name was changed to Petroleum Recovery Institute (PRI). PRI was acquired by ARC in 1999 and later merged into ARC's (now InnoTech's) Energy Division.

**Petroleum Recovery Research
Institute**   (PRRI) *See* Petroleum Recovery Institute.

POP   Persistent organic pollutant.

**Prairie Agricultural Machinery
Institute**   (PAMI) A Humboldt, Saskatchewan-based, not-for-profit, research and technology organization (RTO) created in 1975. An agricultural industry, Saskatchewan and Manitoba government partnership, PAMI focuses on applied research, development, testing, and evaluation related to farm machinery.

PRI   *See* Petroleum Recovery Institute.

PRO   Provincial Research Organization. Provincially owned RTOs in Canada are sometimes referred to as provincial research councils or Provincial Research Organizations (PROs). *See* Research and Technology Organization.

Production Transfer
Pilot Plant   (PTPP) An RPC pilot plant and experimental manufacturing facility. This was building #5, for RPC, and opened in July 1975.

PRRI   *See* Petroleum Recovery Institute.

PTPP   *See* Production Transfer Pilot Plant.

R&D   Research and development, this acronym is frequently meant to refer specifically to applied research and development.

QA   Quality Assurance.

RCA   Research Council of Alberta. *See* InnoTech.

RCS   Research Council of Saskatchewan. *See* Saskatchewan Research Council.

RDC   *See* Research and Development Corp.

Research and Development
Corp.   (RDC) The 10th provincial research council established in Canada, it was created in Newfoundland and Labrador as the Research and Development Council in 2009 and later renamed Research and Development Corp. (RDC). It was wound-up in 2018.

Research and Productivity
Council   (RPC) The sixth provincial research council to be established in Canada, it was created as the New Brunswick Research and Productivity Council in 1962 and later renamed Research and Productivity Council.

Research and Technology

Organization   (RTO) A government-owned corporation/agency or private not-for-profit company that is primarily focused on developing and deploying practical technologies that address commercial marketplace problems or opportunities. RTOs generally span multiple sectors of an economy. RTOs differ from academia, mainstream government, for-profit companies, and even from mainstream not-for-profit companies. They have "public good" missions and are not primarily profit-driven. They work in the public interest but on market-pull issues, generally using a businesslike approach, and frequently with a secondary, socio-environmental agenda. One of the most important functions of an RTO is to help business enterprises access, absorb, adapt, deploy, and exploit new technologies in order to enhance businesses' ability to innovate and, therefore, their competitiveness and sustainability. Also termed Applied Research Organization.

Research Council

of Alberta   (RCA) *See* InnoTech.

Research Council

of Ontario   *See* Ontario Research Foundation.

Research Council of

Saskatchewan   (RCS) *See* Saskatchewan Research Council.

RPC   See Research and Productivity Council.

RTO   *See* Research and Technology Organization.

Saskatchewan Research

Council   (SRC) The third provincial research council to be established in Canada, it was created as the Research Council of Saskatchewan (RCS) 1930, later collapsed, and later re-established as the Saskatchewan Research Council in 1947.

SCC   Standards Council of Canada.

Science and Industrial Research Council

 of Alberta   (SIRCA) *See* Alberta Innovates – Technology Futures.

SIRCA Science and Industrial Research Council of Alberta. *See* Alberta Innovates – Technology Futures.

Small- and Medium-Sized Enterprises (SMEs) Business enterprises that are smaller than a specified number of employees and/or have annual revenues of less than a specified value, as distinguished from large-size enterprises. RPC has consistently defined SMEs as enterprises having fewer than 200 persons.

SMEs *See* Small- and Medium-Sized Enterprises.

SRC *See* Saskatchewan Research Council.

Technical Information Service (TIS) An NRC program aimed at assisting industry by making appropriate research and technical information available to it. This program was delivered by RPC, for NRC, in New Brunswick (beginning in 1963) and Prince Edward Island (beginning in 1964) and continuing until it was superseded by NRC's broader Industrial Research Assistance Program (IRAP) Program in 1981. IRAP focuses on providing a range of technical services to small- and medium-sized enterprises (SMEs) in Canada.

TIS *See* Technical Information Service.

TQM Total Quality Management.

UFFI Urea formaldehyde foam insulation.

UNB University of New Brunswick.

Vineland Research and Innovation Centre One of Canada's first Research and Technology Organizations. It was created in 1906 as the Horticultural Experiment Station "Jordan Harbour" (sometimes called Vineland Research Station), and became the Horticultural Research Institute of Ontario (HRIO) in 1966. In 1997, HRIO was transferred to the University of Guelph, and in 2009 it was spun-out as the not-for-profit Vineland Research and Innovation Centre.

**Vineland Research
Station**   *See* Vineland Research and Innovation Centre.

9 SUMMARY

Research and Development in New Brunswick.
A History of the Research and Productivity Council

Created by the Government of New Brunswick in 1962, the Research and Productivity Council (RPC) quickly took its place among Canada's provincial research organizations. RPC's mission would be to bring organized research, development, and technological innovation to New Brunswick and, specifically, to *"stimulate and expedite continuing improvements in productive efficiency and expansion in various sectors of New Brunswick's economy."* In carrying out its mandate, RPC has helped industry develop and advance, supplying everything from analysis and testing, to expert assistance with the full product/process development cycle – from concept to commercialization. Although not among the first of Canada's RTOs to be established, RPC has evolved to become one of the most enduring, and one of the nation's leaders in the modern approach to technological innovation. This book traces the evolution of RPC through several distinct eras comprising its *Building Years (1962-1969)*, its *Growing Years* (1970-1983), *Commercial Years* (1983-2004), and *Entrepreneurial Years* (2004-2020).

Print ISBN: 978-1-7770816-0-7
ePub ISBN: 978-1-7770816-1-4

10 ABOUT THE AUTHORS

Eric Cook and Dr. Laurier Schramm have been professional colleagues, collaborators, and friends for over 15 years. They first met in December 2004, during the early days of the formation of an alliance of Canada's RTOs named Innoventures Canada Inc. (I-CAN, see Appendix 8.10). As the CEOs of RPC and SRC, respectively, they quickly found that, although their organizations, operations, and business focus were 3,000 km apart, their challenges, opportunities, successes, and sense of mission were remarkably similar. Although originally trained in different disciplines, they also share an interest in the transitions of knowledge and understanding from science to engineering, Eric, being an engineer that leans toward science, and Laurie, a scientist who leans toward engineering, their interests met in the middle. None of these things have changed to the present day.

Over the years, Laurie and Eric also discovered a shared interest in the leadership and evolution of RTOs, with their unusual focus on 'public good' mission over the more conventional 'revenue and profits' model of commercial businesses. For much of the past 15 years, they have worked to support I-CAN, both having served as long-standing members of the Board of Directors, and each having served as I-CAN's Board Chair. Their collaborations have taken them from 'coast-to-coast-to-coast' in Canada, from St. John's, NL, to Victoria, BC, to Cambridge Bay, NT (500 km north of the Arctic Circle).

Dr. Laurier Schramm, Ph.D., P.Chem., C.Dir., FCIC, FCAE is an industrial scientist with 40 years of R&D experience spanning the industry, not-for-profit, university, and government sectors. He served as President and CEO of the Saskatchewan Research Council (SRC) for 18 years, and was Vice-President with the Alberta Research Council, and President and CEO

of the Petroleum Recovery Institute. Laurie has also served on many expert advisory panels and Boards, is co-founder of Innoventures Canada Inc., and co-founder of Canada's Innovation School™. His interests include technological innovation, management and leadership, colloid & interface science, and nanotechnology. He holds 17 patents and has published over 16 books and over 400 other publications and proprietary reports, and has received national scientific and engineering awards for his work.

Eric Cook, MBA, P.Eng., ICD.D. is Executive Director and CEO of the New Brunswick Research and Productivity Council, and was previously General Manager of COM DEV Wireless. His career includes over 25 years of leadership in the application of science and technology in sectors including advanced manufacturing, space science, aerospace, and wireless

communications. His training and background in science, engineering, and business administration provides a combination that has served him well in seeking science-based solutions for business problems and opportunities. Eric has been an active board member for organizations including Genome Canada, the New Brunswick Innovation Foundation, Innoventures Canada Inc., the Center for Nuclear Energy Research, and the New Brunswick Environmental Industries Association. He has also served on a number of expert panels and committees.

11 REFERENCES

1.   *Fredericton Daily Gleaner*, "New Productivity Building Opened," May 21, 1965, pp. 1,2,16.

2.   Dugas, G., "Research, Productivity Council Serves N.B.," *Fredericton Daily Gleaner*, June 22, 1979, pp. 18-19.

3.   RPC, "Annual Report 2002-2003," New Brunswick Research and Productivity Council, Fredericton, 2003.

4.   Palmer, E.F., *Horticultural Experiment Station and Products Laboratory: The First Fifty Years, 1906 to 1956*, Ontario Department of Agriculture, Toronto, 1956 (available from The Atrium, University of Guelph, at: https://atrium.lib.uoguelph.ca/xmlui/handle/10214/3076).

5.   Loughton, A.; Chudyk, R.V.; Wanner, J.A. (Eds.), *Celebrating A Century of Success 1906 – 2006*, Agricultural Experimental Station, Vineland, University of Guelph, Guelph, Ontario, 2006.

6.   Schramm, L.L., *Research and Development on the Prairies. A History of the Saskatchewan Research Council*, Saskatchewan Research Council, Saskatoon and Amazon.com, 2017.

7.   Holland, M., *Industrial Explorers*, Harper & Bros., New York, 1928.

8.   Holland, M., "Research, Science and Invention," In *A Century of Industrial Progress*, Wile, F.W. (Ed.), Doubleday: N.Y., 1928, pp. 312-334.

9.   Schramm, L.L., *Technological Innovation – An Introduction*, de Gruyter, Berlin, *2018*.

10. Eggleston, W., "*National Research in Canada, The NRC 1916 – 1966,*", Clarke, Irwin & Co., Toronto, 1978.

11. Thistle, M., "*The Inner Ring. The Early History of the National Research Council of Canada*, University of Toronto Press, Toronto, 1966.

12. Phillipson, D.J.C., "The National Research Council of Canada: Its Historiography, its Chronology, its Bibliography," *Scientia Canadensis*, **1991**, *15(2)*, 177-193.

13. The Advisory Council of Scientific and Industrial Research of Alberta, "First Annual Report," King's Printer, Edmonton, 1921.

14. The Research Council of Alberta, "Thirty-First Annual Report," King's Printer, Edmonton, 1951.

15. The Research Council of Alberta, "Annual Report, 1943," King's Printer, Edmonton, 1945.

16. Research Council of Alberta, "Thirty-Sixth Annual Report," Queen's Printer, Edmonton, 1956.

17. Aronovitch, D. (Ed.), "Ontario Research Foundation," In *The Canadian Encyclopedia*, 2013, www.thecanadianencyclopedia.ca/en/article/ontario-research-foundation.

18. Le Roy, D.J.; Dufour, P., "Partners in Industrial Strategy. The Special Role of the Provincial Research Organizations," Background Study for the Science Council of Canada, Special Study 51, Supply and Services Canada, Ottawa, Nov. 1983.

19. ORTECH, "Annual Report 1990," ORTECH Corp., Mississauga, 1991.

20. Warren, T.E., "A Brief History of the Saskatchewan Research Council. 1914 – 1972," Report, Saskatchewan Research Council, Saskatoon, 1972.

21. Saskatchewan Archives Board, *Personal Communication*, 24 November 2011.

22. Government of Saskatchewan, *"The Research Council Act,"* 1947, c.114; as amended by The Revised Statutes of Saskatchewan, 1978 (Supplement), c.60; and the Statutes of Saskatchewan, 1983-84, c.34; 1988-89, c.22; 1991, c.T-1.1; 1994, c.45; 2000, c.23; and 2014, c.E-13.1; Queen's Printer, Regina, 2014.

23. Aronovitch, D. (Ed.), "British Columbia Research Council," In *The Canadian Encyclopedia*, 2013, www.thecanadianencyclopedia.ca/en/article/british-columbia-research-council.

24. Aronovitch, D. (Ed.), "Nova Scotia Research Foundation Corporation," In *The Canadian Encyclopedia*, 2013,www.thecanadianencyclopedia.ca/en/article/nova-scotia-research-foundation-corporation.

25. BC Research, "1990-1991 Annual Report, British Columbia Research Corp., Vancouver, 1991.

26. InNOVAcorp, "Annual Report 1995-1996," Nova Scotia Innovation Corp., Halifax, 1996.

27. Wilson, A.H., "Research Councils in the Provinces: A Canadian Resource," Background Study for the Science Council of Canada, Special Study 19, Information Canada, Ottawa, June 1971.

28. Government of New Brunswick, *"An Act to Establish a Research and Productivity Council,"* 1961-62, c.46, Queen's Printer, Fredericton, April 13, 1962.

29. Steacie, E.W.R., "The Government Laboratory and Canadian Research," Presented at Symposium on Research and Opening Ceremonies for SRC's new building, 1 October 1958, from Saskatchewan Research Council historical files, Saskatoon.

30. *Moncton Transcript*, "Motion will Seek Resignation One of Commissioners," April 11, 1962, p.2.

31. *Moncton Daily Times*, "Public Works Minister… To Investigate Alleged Firing," June 20, 1963, p. 3.

32. RPC, "First Annual Report 1962-63," New Brunswick Research and Productivity Council, Fredericton, 1963.

33. *Moncton Transcript*, "$35,000 in Grants to University Faculty," Nov. 10, 1962, p. 10.

34. *Saint John Telegraph Journal*, "Managers to get Training Program," April 13, 1963, p. 17.

35. *Moncton Transcript*, "One-Day Seminar is Held in Bathurst," Dec. 18, 1963, p.17.

36. *Saint John Telegraph Journal*, "Management Training Program," Advertisement, April 23, 1963, p. 17.

37. *Fredericton Daily Gleaner*, "For Manufacturers Only," Advertisement, Sept. 5, 1964, p. 35.

38. RPC, "Second Annual Report 1963-64," New Brunswick Research and Productivity Council, Fredericton, 1964.

39. *Fredericton Daily Gleaner*, "RPC Executive Director Retiring this Month," September 17, 1983, p. 13.

40. *Moncton Daily Times*, "Theirs is a Challenging Task," Editorial, Oct. 8, 1963, p. 4.

41. Bursill, C., "Executive Director's Confidential Report to the New Brunswick Research and Productivity Council Members for the Month of January, 1965," New Brunswick Research and Productivity Council, Fredericton, 1965.

42. Bursill, C., "Science and Industry," Letters, *The Economist*, Jan 16, 1965, p. 177.

43. RPC, "Third Annual Report 1964-65," New Brunswick Research and Productivity Council, Fredericton, 1965.

44. *Saint John Telegraph Journal*, "Premier Stresses Need for Research," Nov. 22, 1963, p. 2.

45. RPC, "Catalog of Research Resources and Activities, New Brunswick, 1963," Proc. Research Inventory Conference, Nov. 21-22, 1963, New Brunswick Research and Productivity Council, Fredericton.

46. *Fredericton Daily Gleaner*, "Research Inventory Conference," Editorial, Nov. 29, 1963, p. 4.

47. *Moncton Daily Times*, "To Assist Forward March of Industry in N.B.," Editorial, Aug. 5, 1964, p. 4.

48. RPC, "Fifth Annual Report 1966-1967," New Brunswick Research and Productivity Council, Fredericton, 1967.

49. RPC, "Sixth Annual Report 1967-1968," New Brunswick Research and Productivity Council, Fredericton, 1968.

50. *Saint John Telegraph Journal*, "Building Planned," May 21, 1965, p. 33.

51. *Moncton Transcript*, "Direct Help," Feb. 16, 1965, p. 10.

52. Bursill, C., "Executive Director's Confidential Report to the New Brunswick Research and Productivity Council Members for the Month of March, 1964," New Brunswick Research and Productivity Council, Fredericton, 1964.

53. Bursill, C., "Executive Director's Confidential Report to the New Brunswick Research and Productivity Council Members for the Month of March, 1965," New Brunswick Research and Productivity Council, Fredericton, 1965.

54. Bursill, C., "Executive Director's Confidential Report to the New Brunswick Research and Productivity Council Members for the Month of April 1967," New Brunswick Research and Productivity Council, Fredericton, 1967.

55. RPC, "Fourth Annual Report 1965-1966," New Brunswick Research and Productivity Council, Fredericton, 1966.

56. *Moncton Transcript*, "ADB Grant for Research, Productivity Council," Feb. 20, 1968, p. 11.

57. CBC, "The Formality of Celebrating a Structure is With Us, and With Us Again," CBC Radio Special coverage of the grand opening of RPC's building #4, June 5, 1968.

58. RPC, "Seventh Annual Report 1968-1969," New Brunswick Research and Productivity Council, Fredericton, 1969.

59. Bursill, C., "Executive Director's Confidential Report to the New Brunswick Research and Productivity Council Members for the Month of Sept. 1965," New Brunswick Research and Productivity Council, Fredericton, 1965.

60. Bursill, C., "Executive Director's Confidential Report to the New Brunswick Research and Productivity Council Members for the Month of Nov. 1964," New Brunswick Research and Productivity Council, Fredericton, 1964.

61. RPC, "Eighth Annual Report 1969-1970," New Brunswick Research and Productivity Council, Fredericton, 1970.

62. Bursill, C., "Executive Director's Confidential Report to the New Brunswick Research and Productivity Council Members for the Month of August 1965," New Brunswick Research and Productivity Council, Fredericton, 1965.

63. Bursill, C., "Executive Director's Confidential Report to the New Brunswick Research and Productivity Council Members for the Month of Sept. 1966," New Brunswick Research and Productivity Council, Fredericton, 1966.

64. Bursill, C., "Executive Director's Confidential Report to the New Brunswick Research and Productivity Council Members for the Month of Jan. 1968," New Brunswick Research and Productivity Council, Fredericton, 1968.

65. Bursill, C., "Executive Director's Confidential Report to the New Brunswick Research and Productivity Council Members for the Month of Oct. 1969," New Brunswick Research and Productivity Council, Fredericton, 1969.

66. McLeod, D.A., "Log skidding cone," U.S. Patent 3,926,410, December 16, 1975.

67. Bursill, C., "Executive Director's Confidential Report to the New Brunswick Research and Productivity Council Members for the Month of July 1967," New Brunswick Research and Productivity Council, Fredericton, 1967.

68. Steen, M., "Locally Made Movies Assist N.B. Firms," *Fredericton Daily Gleaner*, Aug. 29, 1966, p. 9.

69. Bursill, C., "Executive Director's Confidential Report to the New Brunswick Research and Productivity Council Members for the Month of July 1969," New Brunswick Research and Productivity Council, Fredericton, 1969.

70. *Saint John Telegraph Journal*, "New Brunswick Research and Productivity Council," Advertisement, April 6, 1966, p. 9.

71. Gibbs, D., "Chemistry – Food Science Department," *Fredericton Daily Gleaner*, Sept. 14, 1970, p. 3.

72. RPC, "RPC Motivates," Promotional brochure, New Brunswick Research and Productivity Council, Fredericton, 1968, 10 pp.

73. Bursill, C., "Executive Director's Confidential Report to the New Brunswick Research and Productivity Council Members for the Month of June 1970," New Brunswick Research and Productivity Council, Fredericton, 1970.

74. Bursill, C., "Executive Director's Confidential Report to the New Brunswick Research and Productivity Council Members for the Month of Aug. 1969," New Brunswick Research and Productivity Council, Fredericton, 1969.

75. Bursill, C., "Executive Director's Confidential Report to the New Brunswick Research and Productivity Council Members for the Month of Sept. 1969," New Brunswick Research and Productivity Council, Fredericton, 1969.

76. Bursill, C., "Executive Director's Confidential Report to the New Brunswick Research and Productivity Council Members for the Month of Dec. 1969," New Brunswick Research and Productivity Council, Fredericton, 1969.

77. RPC, "Eleventh Annual Report 1972-1973," New Brunswick Research and Productivity Council, Fredericton, 1973.

78. Bursill, C., "Executive Director's Confidential Report to the New Brunswick Research and Productivity Council Members for the Month of Feb. 1972," New Brunswick Research and Productivity Council, Fredericton, 1972.

79. *Fredericton Daily Gleaner*, "Appointed to Head Council," June 8, 1970, p. 16.

80. RPC, "Ninth Annual Report 1970-1971," New Brunswick Research and Productivity Council, Fredericton, 1971.

81. Bursill, C., "Executive Director's Confidential Report to the New Brunswick Research and Productivity Council Members for the Month of May 1973," New Brunswick Research and Productivity Council, Fredericton, 1973.

82. RPC, "Twelfth Annual Report 1973-1974," New Brunswick Research and Productivity Council, Fredericton, 1974.

83. *Moncton Transcript*, "RPC is Specialist Group," June 1, 1970, p. 1.

84. Gibbs, D., "N.B.'s Research Council Solves Industry Problems," *Fredericton Daily Gleaner*, Sept. 12, 1970, p. 10.

85. Gibbs, D., "RPC's Management Experts 'Spin-Off' to Industry," *Fredericton Daily Gleaner*, Sept. 15, 1970, p. 5.

86. Gibbs, D., "National Council Backs 2 RPC Services," *Fredericton Daily Gleaner*, Sept. 16, 1970, p. 14.

87. Gibbs, D., "RPC Unit Conducts Ore Studies," *Fredericton Daily Gleaner*, Sept. 17, 1970, p. 10.

88. Gibbs, D., "RPC Unit Conducts Ore Studies," *Fredericton Daily Gleaner*, Sept. 17, 1970, p. 10.

89. RPC, "A Photo Tribute to Dr. Roy S Boorman on the Occasion of His Retirement," Internal video, New Brunswick Research and Productivity Council, Fredericton, October 13, 1992.

90. RPC, "Tenth Annual Report 1971-1972," New Brunswick Research and Productivity Council, Fredericton, 1972.

91. *Saint John Telegraph Journal*, "'Bricklin' Showing Next Month," Jan. 5, 1974, p. 20.

92. RPC, "Thirteenth Annual Report 1974-1975," New Brunswick Research and Productivity Council, Fredericton, 1975.

93. *Saint John Telegraph Journal*, "Bricklin Owes Money to RPC," Mar. 22, 1975, p. 2.

94. *Saint John Telegraph Journal*, "Living Standard Suggested as Measure of N.B. Growth," June 5, 1975, p. 3.

95. *Saint John Telegraph Journal*, "Higgins Assails PCs," Oct. 18, 1976, p.5.

96. Bricklin Autosport, "History of the Bricklin Car," Avanquest Software, Inc., 2007,
https://web.archive.org/web/20161020033315/http://www.bricklinautosport.com/bricklin_autosport_003.htm

97. https://commons.wikimedia.org/wiki/File:1974_AACA_museum_Bricklin_green_l.jpg

98. *Moncton Transcript*, "Wide Program Outlined," March 6, 1974, p. 2.

99. Bursill, C., "Executive Director's Confidential Report to the New Brunswick Research and Productivity Council Members for the Month of July 1975," New Brunswick Research and Productivity Council, Fredericton, 1975.

100. Bursill, C., "Executive Director's Confidential Report to the New Brunswick Research and Productivity Council Members for the Month of Aug. 1975," New Brunswick Research and Productivity Council, Fredericton, 1975.

101. *Fredericton Daily Gleaner*, "Kasper and Richter Holds Opening Ceremony Today," Feb. 5, 1975, p. 35.

102. RPC, "Fourteenth Annual Report 1975-1976," New Brunswick Research and Productivity Council, Fredericton, 1976.

103. Bursill, C., "Executive Director's Confidential Report to the New Brunswick Research and Productivity Council Members for the Month of Sept. 1976," New Brunswick Research and Productivity Council, Fredericton, 1976.

104. Bursill, C., "Executive Director's Confidential Report to the New Brunswick Research and Productivity Council Members for the Month of Nov. 1976," New Brunswick Research and Productivity Council, Fredericton, 1976.

105. Bursill, C., "Executive Director's Confidential Report to the New Brunswick Research and Productivity Council Members for the Month of Jan. 1977," New Brunswick Research and Productivity Council, Fredericton, 1977.

106. *Fredericton Daily Gleaner*, "Firm Behind Canadian Merchandise," Advertisement, Sept. 13, 1977, p. 13.

107. *Fredericton Daily Gleaner*, "Galleon Ware," Advertisement, Dec. 10, 1976, p. 19.

108. Bursill, C., "Executive Director's Confidential Report to the New Brunswick Research and Productivity Council Members for the Month of March 1977," New Brunswick Research and Productivity Council, Fredericton, 1977.

109. Bursill, C., "Executive Director's Confidential Report to the New Brunswick Research and Productivity Council Members for the Months of April through September 1977," New Brunswick Research and Productivity Council, Fredericton, 1977.

110. Orser, F., "Local Company Competes for Share of Tableware Market," *Fredericton Daily Gleaner*, April 12, 1978, p. 21.

111. See, for example: https://www.etsy.com/ca/market/galleon_ware; https://www.kijiji.ca/v-art-collectibles/; and https://www.ebay.ca › Collectibles › Decorative Collectibles › Mugs, Cups.

112. *Fredericton Daily Gleaner*, "Chamber of Commerce Driving Force Behind Development," Feb. 13, 1978, p. 27.

113. *Fredericton Daily Gleaner*, "Restricted Use of Oil in Region Forecast – Greater Energy Consciousness at Home Used," Feb. 14, 1978, p. 3.

114. *Moncton Transcript*, "N.B. has Energy Bus Program," Oct. 29, 1979, p. 13.

115. RPC, "1986-87 Annual Report," New Brunswick Research and Productivity Council, Fredericton, 1987.

116. *Fredericton Daily Gleaner*, "Do You Fit in This Picture?" Advertisement, Dec. 3, 1981, p. 20.

117. *Fredericton Daily Gleaner*, "This Bus Can Save Your Company Money!" Advertisement, Aug. 2, 1979, p. 50.

118. *Saint John Telegraph Journal*, "$300,000 Let for Energy Projects," Feb. 8, 1983, p. 7.

119. *Saint John Telegraph Journal*, "RPC Research and Productivity Council," Advertisement, June 23, 1983, p. 33.

120. *Fredericton Daily Gleaner*, "Businessmen Told Standards for Nuclear Projects," March 14, 1978, p. 2.

121. *Fredericton Daily Gleaner*, "RPC Uses Scientific Technology to Aid Industry," June 22, 1979, p. 18.

122. *Saint John Telegraph Journal*, "Point Lepreau Project Generators being Retubed," Oct. 11, 1980, p. 3.

123. RPC, "Twentieth Annual Report 1981-1982," New Brunswick Research and Productivity Council, Fredericton, 1982.

124. RPC, "Fifteenth Annual Report 1976-1977," New Brunswick Research and Productivity Council, Fredericton, 1977.

125. RPC, "Sixteenth Annual Report 1977-1978," New Brunswick Research and Productivity Council, Fredericton, 1978.

126. Science Council, "Profiles of the Provincial Research Organizations," Science Council of Canada, Ottawa, 1981.

127. New Brunswick, "Elmtree Limestone," and "Petit Rocher (Elmtree, LaPlante) Limestone," Industrial Minerals Summary Data, Energy and Resource Development, Government of New Brunswick, https://www1.gnb.ca/0078/GeoscienceDatabase/IndustrialMinerals/qryInd MinSummary-e.asp?Num=827, and https://www1.gnb.ca/0078/GeoscienceDatabase/IndustrialMinerals/qryInd MinSummary-e.asp?Num=451, respectively.

128. Tetra Tech, "Preliminary Economic Assessment on the Woodstock Manganese Property, New Brunswick Canada," Report for Canadian Manganese Company Inc., Tetra Tech, Toronto, July 10, 2014.

129. Gilders, R.D., "Geology of Woodstock Manganese Claim group for R.S. Boorman," New Brunswick, Assessment Report 470389 filed with the N.B. Department of Natural Resources, 1976.

130. Gilders, R.D., "Geophysics of Woodstock Manganese Claim Group (19 claims) for R.S. Boorman (Minuvar Ltd.)," New Brunswick, Assessment Report 472328 filed with the N.B. Department of Natural Resources, 1978.

131. RPC, "Nineteenth Annual Report 1980-1981," New Brunswick Research and Productivity Council, Fredericton, 1981.

132. *Fredericton Daily Gleaner*, "NBRPC Wants Smelting Pilot Plant. New Technology Would Up Metal Recovery Rate," April 26, 1983, p. 3.

133. *Moncton Transcript*, "New Ore Refining Process May Have Major Impact for N.B.," Oct. 17, 1978, p. 44.

134. RPC, "Eighteenth Annual Report 1979-1980," New Brunswick Research and Productivity Council, Fredericton, 1980.

135. *Fredericton Daily Gleaner*, "N.B. on Brink of Mining Boom. Pilot Project Possible as Early as February," Nov. 21, 1979, p. 1.

136. *Fredericton Daily Gleaner*, "Ore Extraction Plant Planned," Aug. 6, 1983, p. 5.

137. ERS, "First Annual Report," Enhanced Recovery Systems Ltd., Chatham, 1984.

138. RPC, "Seventeenth Annual Report 1978-1979," New Brunswick Research and Productivity Council, Fredericton, 1979.

139. Macauley, G., Powell, T.G., Snowdon, L.R., "Some Geological Considerations for the Economic Evaluation of Canadian Oil Shale Deposits," In *Sixteenth Oil Shale Symposium Proceedings*, Gary, J.H. (Ed.), Colorado School of Mines Press, Golden, 1983, pp. 1-13.

140. *Saint John Telegraph Journal*, "RPC Looks at Means to Use N.B. Shales," Oct. 2, 1982, p. 5.

141. Salib, P.F., Barua, S.K., Sy, O., Gemmell, D.E., Furimsky, E., "Demonstration of an Integrated Fluidized Bed Combustor and Oil Shale Retort System, in *Proc. Eastern Oil Shale Symposium*, Pettit, R. (Ed.), 1985, *IMMR84/124.*

142. Sainz, F.A., Warfe, W.A., Douglas, M.A., Day, R.F., Anders, R., "Chatham Circulating Fluidized Bed Demonstration Project," in Basu, P. (Ed.), *Circulating Fluidized Bed Technology: Proceedings of the First International Conference on Circulating Fluidized Beds*, Elsevier, Amsterdam, 1986, pp. 349-362.

143. *Saint John Telegraph Journal*, "RPC Research and Productivity Council," Advertisement, Feb. 15, 1983, p. 32.

144. O'Donoghue, P. (York Herald, College of Arms, London, England), *Personal communication* to L.L. Schramm, dated August 26, 2019, confirming that "*Arms, Crest and Supporters, and a Badge, were granted to New Brunswick Research and Productivity Council, by Letters Patent of the Kings of Arms dated 20 September 1978.*"

145. RPC, "Twenty-First Annual Report 1982-1983," New Brunswick Research and Productivity Council, Fredericton, 1983.

146. Boorman, R.S., "Executive Director's Confidential Report to the New Brunswick Research and Productivity Council Members for the Month of Oct. 1983," New Brunswick Research and Productivity Council, Fredericton, 1983.

147. Witcomb, B., "RPC Entering Most Critical Stage. Dr. Boorman Plans Different Direction," *Fredericton Daily Gleaner*, Oct. 15, 1983, p. 15.

148. RPC, "Twenty-Second Annual Report 1983-1984," New Brunswick Research and Productivity Council, Fredericton, 1984.

149. RPC, "Twenty-Third Annual Report 1984-1985," New Brunswick Research and Productivity Council, Fredericton, 1985.

150. RPC, "1987-1988 Annual Report," New Brunswick Research and Productivity Council, Fredericton, 1988.

151. RPC, "1988-89 Annual Report," New Brunswick Research and Productivity Council, Fredericton, 1989.

152. *Moncton Times Transcript*, "RPC Receives $19.9M from ACOA," News Briefs, Sept. 27, 1988, pp. 33.

153. Boorman, R.S., "Executive Director's Confidential Report to the New Brunswick Research and Productivity Council Members for the Month of Dec. 1986," New Brunswick Research and Productivity Council, Fredericton, 1986.

154. *New Brunswick Telegraph Journal*, "Legislation Intended to Encourage Business Donations to Food Banks," Apr. 17, 1992, p. 18.

155. RPC, *Alert*, New Brunswick Research and Productivity Council, Fredericton, 3(3) Fall 1987.

156. RPC, *Alert*, New Brunswick Research and Productivity Council, Fredericton, 6(1) Spring 1991.

157. RPC, "1989-90 Annual Report," New Brunswick Research and Productivity Council, Fredericton, 1990.

158. RPC, "1990-91 Annual Report," New Brunswick Research and Productivity Council, Fredericton, 1991.

159. *Fredericton Daily Gleaner,* "RPC Undergoes Major $3-Million Expansion to Facility," Feb. 14, 1991, p. 45.

160. RPC, "1985-86 Annual Report," New Brunswick Research and Productivity Council, Fredericton, 1986.

161. RPC, *Inside Out*, New Brunswick Research and Productivity Council, Fredericton, Oct. 1990, pp. 1-2.

162. RPC, *Alert*, New Brunswick Research and Productivity Council, Fredericton, 3(4) Winter 1988.

163. *Saint John Telegraph Journal*, "Chatham-Area Plant Hopes to get More Out of Metals," May 14, 1985, p. 13.

164. *Saint John Telegraph Journal*, "Chatham Demonstration Project Shuts Down Because of Lack of Funding," Oct. 4, 1986, p. 40.

165. *Saint John Telegraph Journal*, "NB Power Plans to Make Chatham Plant into Showcase," June 21, 1985, p. 6.

166. *New Brunswick Telegraph Journal*, "Bringing 'Ideas to Life' is Work of RPC," Feb. 27, 1992, p. 46.

167. RPC, *Alert*, New Brunswick Research and Productivity Council, Fredericton, Winter 1990.

168. *Fredericton Daily Gleaner,* "Fish Diagnostic Kit," Feb. 12, 1991, p. 22.

169. *New Brunswick Telegraph Journal*, "Mending the Net when the Fish are Gone," Jan. 7, 1998, p. 9.

170. Morgan, S., "Cutting Fish Plant Effluent," *Fredericton Daily Gleaner*, Sept. 15, 1998, p. 42.

171. Morgan, S., "Scientists Isolate Virus Killing Atlantic Salmon," *Fredericton Daily Gleaner*, Sept. 11, 1997, p. 3.

172. Young, D., "The Hunt for the Phantom Fish-Killer. How a Mysterious Virus Exposed Mistrust and Ill-Preparedness Among Fish Farmers, the Province and Scientists and Nearly Crippled a $120-million-a-year Industry," *New Brunswick Telegraph Journal*, Aug. 22, 1998, p. 90-96.

173. LeBlanc, R., "N.B.-Developed Salmon Vaccine Promises to Provide Boost to Worldwide Aquaculture," *New Brunswick Telegraph Journal*, Aug. 30, 2001, pp. 1,8.

174. Senate of Canada, Remarks by Dr. Ben Forward related to the regulation of aquaculture, current challenges and future prospects for the industry in Canada, In *Proc., Standing Senate Committee on Fisheries and Oceans,* "Issue 14 – Evidence," Senate of Canada, Ottawa, November 20, 2014, https://sencanada.ca/en/Content/Sen/Committee/412/POFO/14ev-51765-e

175. RPC, "Annual Report 2000-2001," New Brunswick Research and Productivity Council, Fredericton, 2001.

176. *Fredericton Daily Gleaner,* "Research: A Window on the Future," July 8, 2004, pp. 49-50.

177. Ritchie, R., "'CSI' Science Helps Improve Genetic Diversity of Fish," *New Brunswick Telegraph Journal*, June 4, 2004, p. 52.

178. *New Brunswick Telegraph Journal*, "The Little Bugs That Could," Sept. 25, 2004, p. 9.

179. *Fredericton Daily Gleaner,* "UNB Manufacturing Technology Centre. Business and Revenue Quintupled in 5 Years," Feb. 11, 1988, p. 43.

180. *Fredericton Daily Gleaner,* "Technology Transfer Agreement to be Signed," Mar. 21, 1984, p. 10.

181. *Saint John Telegraph Journal*, "Computer Training Agreement Signed," Mar. 22, 1984, p. 26.

182. *Fredericton Daily Gleaner,* "RPC Provides Technological Excellence," Feb. 21, 1985, p. 43.

183. *Saint John Telegraph Journal*, "Company Given Grant," Feb. 21, 1989, p. 13.

184. *Fredericton Daily Gleaner,* "Several City Projects Receive Assistance from ACOA," Feb. 27, 1989, p. 21.

185. *Saint John Telegraph Journal*, "MTC/CADMI Inc.," advertisement, June 18, 1990, p. 35.

186. *Fredericton Daily Gleaner,* "Research Council Develops Ideas for N.B. Industry," Feb. 12, 1987, p. 44.

187. RPC, "Research and Productivity Council," Promotional brochure, New Brunswick Research and Productivity Council, Fredericton, *circa* 1983, 12 pp.

188. RPC, *Alert*, New Brunswick Research and Productivity Council, Fredericton, 2(4) November 1986.

189. RPC, "Meeting the Challenges of Industry," Promotional video, New Brunswick Research and Productivity Council, Fredericton, 1986.

190. RPC, *Inside Out*, New Brunswick Research and Productivity Council, Fredericton, Aug. 1990, pp. 1-2.

191. Meagher, D., "Believed to Involve Coil of Lithium. RPC Fire Under Investigation," *Fredericton Daily Gleaner*, July 25, 1990, pp. 17,19.

192. *Fredericton Daily Gleaner*, "RPC Meeting Challenges of Today," Feb. 2, 1986, p. 48.

193. *Saint John Telegraph Journal*, "Council Benefits Many in Province," Feb. 27, 1986, p. 83.

194. *Saint John Telegraph Journal*, "It's a Technology Centre for N.B. Industry," June 5, 1984, p. 26.

195. CBC, "The Tainted Star-Kist Tuna Scandal," CBC Digital Archives, Canadian Broadcasting Co., 2018, https://www.cbc.ca/archives/entry/the-tainted-star-kist-tuna-scandal.

196. *Moncton Times Transcript*, "Fraser Overrules Inspectors; Gives OK to Rancid Tuna, Program Says," Sept. 18, 1985, pp. 1,31.

197. *Saint John Telegraph Journal*, "CBC Charges: Fraser Cleared Suspect Tuna at Hatfield's Urging," Sept. 18, 1985, p. 1.

198. *Saint John Telegraph Journal*, "N.B. Stores Pull Star-Kist Tuna," Sept. 19, 1985, pp. 1-2.

199. Dunsmuir, H., "Reaction Scanty," *Fredericton Daily Gleaner*, Sept. 18, 1985, p. 1.

200. *Fredericton Daily Gleaner*, "RPC Explains Tuna Role," Sept. 21, 1985, p. 12.

201. Estill, M., "Premier: Right Action Taken," *Fredericton Daily Gleaner*, Sept. 19, 1985, p. 1.

202. MacDonald, A., "Tuna Released Before Tests Were Finished," *Moncton Times Transcript*, Sept. 21, 1985, p. 3.

203. Camp, D., "A Contrary View on Tuna Affair: Inspectors, not Fraser, at Fault," *Toronto Star*, September 24, 1985, p. A10.

204. MacDonald, A., "Objectivity of Inspectors at Issue - Hatfield," *Moncton Times Transcript*, Sept. 19, 1985, p. 1.

205. Maloney, J., "Insect Management in Canada's Forest Sector: Strengthening National Cooperation Against Current and Future Outbreaks," Report of the Standing Committee on Natural Resources, House of Commons, Ottawa, February 2019.

206. RPC, *Alert*, New Brunswick Research and Productivity Council, Fredericton, Winter/Spring 1996.

207. *Fredericton Daily Gleaner*, "Budworm Conference," and several other articles, Jan. 25, 1990, p. 5.

208. RPC, *Visions*, New Brunswick Research and Productivity Council, Fredericton, Winter/Spring 1996.

209. RPC, *Alert*, first issue, New Brunswick Research and Productivity Council, Fredericton, 1(1) December 1984.

210. RPC, "Annual Report 1999-2000," New Brunswick Research and Productivity Council, Fredericton, 2000.

211. RPC, *Alert*, New Brunswick Research and Productivity Council, Fredericton, 5(4) Winter 1990.

212. RPC, *Inside Out*, New Brunswick Research and Productivity Council, Fredericton, Nov. 1990.

213. RPC, "Annual Report 2003-2004," New Brunswick Research and Productivity Council, Fredericton, 2004.

214. *Fredericton Daily Gleaner*, "RPC Has Key Role to Play," Editorial, June 13, 1992, p. 6.

215. RPC, *Alert*, New Brunswick Research and Productivity Council, Fredericton, Summer 1992.

216. *Fredericton Daily Gleaner*, "Nuclear Research Centre Established at UNB," Apr. 29, 1992, p. 26.

217. UNB, "About Us," Centre for Nuclear Energy Research, University of New Brunswick, Fredericton, www.unb.ca/research/cner/about/index.html.

218. CNER, "General Manager," Advertisement, *Saint John Telegraph Journal*, Mar. 19, 2011, p. 47.

219. RPC, *Inside Out*, New Brunswick Research and Productivity Council, Fredericton, July 1993.

220. RPC, *Inside Out*, New Brunswick Research and Productivity Council, Fredericton, July 1992, pp. 1-2,4.

221. RPC, "1992-93 Annual Report," New Brunswick Research and Productivity Council, Fredericton, 1993.

222. RPC, "1996-1997 Annual Report," New Brunswick Research and Productivity Council, Fredericton, 1997.

223. RPC, *Inside Out*, New Brunswick Research and Productivity Council, Fredericton, December 1994.

224. Goguen, G., "Source of Mystery Odour in Lab at Chalmers Remains Unknown," *Saint John Telegraph Journal*, Feb. 14, 1998, p. 3.

225. Cameron, A., "Blocked Drain Source of Hospital Lab's Air Ills," *Saint John Telegraph Journal*, Feb. 21, 1998, p. 5.

226. RPC, "Annual Report 1997-1998," New Brunswick Research and Productivity Council, Fredericton, 1998.

227. Benteau, S., "N.B. Incubator Mall to Get Building," *Fredericton Daily Gleaner*, Mar. 9, 1988, p. 5.

228. ACOA, "Doing Business in the Capital Region," promotional video, ACOA and N.B. Dept. of Commerce and Technology, Fredericton, *circa* 1989.

229. RPC, *Inside Out*, New Brunswick Research and Productivity Council, Fredericton, March 1991, p. 1.

230. *Fredericton Daily Gleaner*, "Incutech Aiding New Firm," Feb. 14, 1991, p. 45.

231. *Saint John Telegraph Journal*, "N.B. Business Incubator Hatches Expansion," May 18, 1995, p. 15.

232. Meagher, D., "Province Closing Two Innovation Agencies," *Fredericton Daily Gleaner*, July 5, 1991, p. 1.

233. Bockus, K., "Incutech an N.B. Success Story," *Fredericton Daily Gleaner*, Feb. 20, 1992, p. 39.

234. Hill, M., "Group Plots its Future," *Fredericton Daily Gleaner*, Feb. 22, 2001, p. 24.

235. RPC, *Inside Out*, New Brunswick Research and Productivity Council, Fredericton, May 1993, pp. 1,4.

236. *New Brunswick Telegraph Journal*, Advertisement, Oct. 27, 1990, p. 24.

237. RPC, "1993-94 Annual Report," New Brunswick Research and Productivity Council, Fredericton, 1994.

238. *Fredericton Daily Gleaner*, "RPC Achieves ISO Registration for Quality System," Advertisement, June 23, 2001, p. 11.

239. RPC, "46th Annual Report 2007-2008," New Brunswick Research and Productivity Council, Fredericton, 2008.

240. Staples, M., "RPC Able to Hang on During Recession," *Fredericton Daily Gleaner*, Mar. 11, 1992, p. 15.

241. RPC, "Annual Report 2001-2002," New Brunswick Research and Productivity Council, Fredericton, 2002.

242. RPC, "Annual Report 2004-2005," New Brunswick Research and Productivity Council, Fredericton, 2005.

243. UNB, "Eric Cook Proves You Don't Need to Leave New Brunswick to Drive Innovation," Engineering Alumni Blog, University of New Brunswick, Fredericton, Feb. 6, 2018, blogs.unb.ca/engineering/2018/02/eric-cook-proves-you-don-t-need-to-leave-new-brunswick-to-drive-innovation.php

244. Progress, "Appointments," RPC advertisement, *Progress Magazine*, **2005**, *12(1)*, p. 50.

245. RPC, "Annual Report 2005-2006," New Brunswick Research and Productivity Council, Fredericton, 2006.

246. RPC, "Annual Report 2006-2007," New Brunswick Research and Productivity Council, Fredericton, 2007.

247. *Fredericton Daily Gleaner*, "NB Salmon DNA Research May Help Open World Markets," Nov. 20, 2006, p. 5.

248. *Saint John Telegraph Journal*, "Six New Brunswick Companies get ACOA Funding," Jan. 22, 2008, p. 8.

249. RPC, "47th Annual Report 2008-2009," New Brunswick Research and Productivity Council, Fredericton, 2009.

250. Babstock, C., "N.B. Tests Stun Guns," *Moncton Times Transcript*, Feb. 11, 2009, p. 1.

251. Mullin, K., "RPC First in N.B. to Offer Relationship Analysis," *Fredericton Daily Gleaner*, Apr. 30, 2009, p. 23.

252. Llewellyn, S., "DNA Testing Not Just for CSI," *Fredericton Daily Gleaner*, Jan. 20, 2011, p. 13.

253. Mullin, K., "Device Detects Fish-Killing Viruses," *Fredericton Daily Gleaner*, Aug. 6, 2009, p. 23.

254. RPC, "49th Annual Report 2010-2011," New Brunswick Research and Productivity Council, Fredericton, 2011.

255. RPC, "48th Annual Report 2009-2010," New Brunswick Research and Productivity Council, Fredericton, 2010.

256. RPC, "Annual Report 2014-2015," New Brunswick Research and Productivity Council, Fredericton, 2015.

257. RPC, "Annual Report 2018-2019," New Brunswick Research and Productivity Council, Fredericton, 2019.

258. Schramm, L.L., *Innovation Technology: A Dictionary*, de Gruyter, Berlin, Germany, 2017.

259. Llewellyn, S., "Centre Expands Air Testing Lab," *Fredericton Daily Gleaner*, Aug. 26, 2010, p. 27.

260. *Fredericton Daily Gleaner*, "Chalmers Hospital Ready to Celebrate 15th Birthday," Sept. 19, 1991, p. 36.

261. RPC, "Annual Report 2011-2012," New Brunswick Research and Productivity Council, Fredericton, 2012.

262. RPC, "Annual Report 2012-2013," New Brunswick Research and Productivity Council, Fredericton, 2013.

263. RPC, "Annual Report 2013-2014," New Brunswick Research and Productivity Council, Fredericton, 2014.

264. Stewart, L., "Firm in Province Leads Science Behind the Smoke in Medical Pot Testing for Several Big Producers," *Saint John Telegraph Journal*, Sept. 10, 2016, pp. D1-D2.

265. RPC, "Annual Report 2017-2018," New Brunswick Research and Productivity Council, Fredericton, 2018.

266. Chilibeck, J., "Provincial Test Labs Ride High on Cannabis," *Saint John Telegraph Journal*, Feb. 5, 2018, pp. A1, A5.

267. Province of New Brunswick, Order in Council OIC2014-300E, Fredericton, August 6, 2014.

268. Province of New Brunswick, Order in Council OIC2016-206E, Fredericton, August 17, 2016.

269. RPC, "Annual Report 2015/2016," New Brunswick Research and Productivity Council, Fredericton, 2016.

270. New Brunswick, "Strategic Program Review Launched," News Release, Executive Council, Government of New Brunswick, Fredericton, January 13, 2015.

271. New Brunswick, "Choices to Move New Brunswick Forward. Strategic Program Review, Government of New Brunswick, Fredericton, November 2015.

272. Province of New Brunswick, An Act Respecting the Consolidation of Certain Laboratories with the New Brunswick Research and Productivity Council, 2017-144, Fredericton, June 19, 2017.

273. RPC, "Annual Report 2016-2017," New Brunswick Research and Productivity Council, Fredericton, 2017.

274. RPC, "Aquatic Toxicology Added to RPC's Scope of Accredited Testing," News Release, New Brunswick Research and Productivity Council, Fredericton, December 14, 2016.

275. Guthrie, L., "World-Class Research is Taking Place on College Hill Road," *Fredericton Daily Gleaner*, Oct. 7, 2016, p. 9.

276. PMO Canada, "Prime Minister Introduces Canada's New Top Scientist," News Release, Prime Minister's Office, Government of Canada, Ottawa, September 26, 2017, https://pm.gc.ca/en/news/news-releases/2017/09/26/prime-minister-introduces-canadas-new-top-scientist.

277. Ontario, "Ontario Names Molly Shoichet the Province's First Chief Scientist," News Release, Office of the Premier, Government of Ontario, Toronto, November 17, 2017, https://news.ontario.ca/opo/en/2017/11/ontario-names-molly-shoichet-the-provinces-first-chief-scientist.html.

278. Marleau, J.N.; Girling, K.D., "Keeping Science's Seat at the Decision-Making Table: Mechanisms to Motivate Policy-Makers to Keep Using Scientific Information in the Age of Disinformation," *FACETS*, **2017**, *2*, 1045–1064.

279. RPC, "Appointment of Dr. Diane Botelho as RPC's First Chief Science Officer," News Release, New Brunswick Research and Productivity Council, Fredericton, November 30, 2018.

280. RPC, "Annual Report 2018-2019," New Brunswick Research and Productivity Council, Fredericton, 2019.

281. *Fredericton Daily Gleaner*, "As it Enters the 90s. Research and Productivity Council has Note of Optimism," Feb. 15, 1990, p. 88.

282. Andrews, E.L., "A Shampoo to Eliminate Odor of Skunk Spray," *The New York Times*, July 8, 1989, p. 32.

283. RPC, "Annual Report 1994-95," New Brunswick Research and Productivity Council, Fredericton, 1995.

284. RPC, "Annual Report 1995-1996," New Brunswick Research and Productivity Council, Fredericton, 1996.

285. Schramm, L.L.; Nyirfa, W.; Grismer, K.; Kramers, J., "Research and Development Impact Assessment for Innovation-Enabling Organizations," *Canadian Public Administration*, **2011**, *54(4)*, 567-581.

286. RPC, *Alert*, New Brunswick Research and Productivity Council, Fredericton, 5(1) 1989.

287. RPC, *Alert*, New Brunswick Research and Productivity Council, Fredericton, Summer 1994.

288. RPC, *Alert*, New Brunswick Research and Productivity Council, Fredericton, Summer 1991.

289. Cook, M., Vincent, S., Ritchie, R., Griffiths, S., "Development of a Strain Typing Assay for Infectious Salmon Anemia Virus (ISAV)," in Proc. Internat. Response to Infectious Salmon Anemia: Prevention, Control, and Eradication, Miller, O., and Cipriano, R.C. Eds.), Tech. Bull. No. 1902, Animal and Plant Health Inspection Service, U.S. Dept. of Agriculture, Washington, 2003, pp. 75-86.

290. Bursill, C., "Executive Director's Confidential Report to the New Brunswick Research and Productivity Council Members for the Month of April, 1966," New Brunswick Research and Productivity Council, Fredericton, 1966.

291. *Fredericton Daily Gleaner*, Cooke Aquaculture Inc. Advertisement, July 8, 2004, p. 16.

292. Clare, S., "Leaders of Innovation in Thriving Industry," *eNBusiness*, **2***(2)*, March 2008, pp. 8-9.

293. House, J., Myles, D.L., "Seventy Years of Service," NB Power, 1990, https://www.nbpower.com/en/about-us/history.

294. Wilson, A.H., "Research Councils in the Provinces: A Canadian Resource," Background Study for the Science Council of Canada, Special Study 19, Information Canada, Ottawa, June, 1971.

295. Bursill, C., "Executive Director's Confidential Report to the New Brunswick Research and Productivity Council Members for the Month of Oct. 1970," New Brunswick Research and Productivity Council, Fredericton, 1970.

296. RPC, *Inside Out*, New Brunswick Research and Productivity Council, Fredericton, April 1990.

297. APRO, "Growing Concerns: Problems and Opportunities of Small and Medium Size Enterprises in Canada," Discussion paper, Association of Provincial Research Organizations, Ottawa, ON, 9 April 1991.

298. SRC, *Annual Report 1991-1992*, Saskatchewan Research Council, Saskatoon, 1992.

299. I-CAN, "Accessing the System," Innoventures Canada Report 2011," Innoventures Canada Inc., Ottawa, 2011.

> *"Growth is the only evidence of life. RPC does and will have life. We must and will continue to grow..."*
>
> Dr. L.W. Shemilt, RPC Chairman,
> June 5, 1968 [57]

"the result validates the deeds"

9 781777 081607